Step On It!

A wild ride through the motor age

TONY DAVIS

BANTAM
SYDNEY AUCKLAND TORONTO NEW YORK LONDON

STEP ON IT!
A BANTAM BOOK
First published in Australia and New Zealand in 2006 by Bantam

Copyright © Tony Davis, 2006
All rights reserved. No part of this publication may be reproduced, stored in a retrieval system, transmitted in any form or by any means, electronic, mechanical, photocopying, recording or otherwise, without the prior written permission of the publisher.

Every effort has been made to identify copyright holders and photographers of pictures in this book. The publishers would be pleased to hear from any copyright holders who have not been acknowledged.

National Library of Australia
Cataloguing-in-Publication Entry

Davis, Tony
　Step on it!: a wild ride through the motor age.
　　ISBN 978 1 86325 528 8.
　　ISBN 1 86325 528 1.
　　1. Automobiles – History. 2. Automobiles – Humor. I. Title.
629.222

Transworld Publishers,
a division of Random House Australia Pty Ltd
20 Alfred Street, Milsons Point, NSW 2061
http://www.randomhouse.com.au
Random House New Zealand Limited
18 Poland Road, Glenfield, Auckland
Transworld Publishers,
a division of The Random House Group Ltd
61–63 Uxbridge Road, Ealing, London W5 5SA
Random House Inc
1745 Broadway, New York, New York 10036

Photographs supplied by Marque Publishing Company, PO Box 1896, Gosford NSW 2250
Photograph of Chuck Berry (p. 52) courtesy Australian Picture Library; Ralph Nader (p. 135) courtesy AAP Image; James Dean (p. 167) courtesy MPTV; Bugatti and Facel Vega (p. 91), Voisin (p. 102), Aerovette and Marzal (p. 109) courtesy Claude Ludi.

Designed by Darian Causby/Highway 51 Design Works
Typeset by Anna Warren, Warren Ventures
Printed and bound by Ligare Book Printer
10 9 8 7 6 5 4 3 2 1

Also by Tony Davis
Lemon! 60 heroic failures of motoring
Extra Lemon! More heroic failures of motoring

The author would like to thank: Carolyn Walsh, Jessica Dettmann, Jude McGee, Alistair Kennedy, Joshua Dowling, Ray Black, Tim Vaughan, Nadine Giusti, Edward Rowe, Anna Warren, Darian Causby, Bob Hall, Barry Lake, Claude Ludi and most of all my father, Pedr Davis, who gave me the automotive interest and much of the education.

Contents

Introduction: Welcome to the future 5

1. Vive la rev 12

2. Turning up the volume 26

3. The drive-in culture 43

4. Australia joins in 57

5. Maxim power 75

6. Sharp curves 90

7. Cars on film 110

8. The boom time 129

9. Off the track 150

10. Addicted to speed 169

11. It seemed like a good idea 186

12. Simply the best 205

INTRODUCTION

Welcome to the future

'Some scientists are not yet ready to acknowledge that atomic-powered automobiles will ever be a reality, or that there is any economic necessity for such transportation. From the standpoint of present-day information there is much to be said for the arguments [for atomic cars].'

That was America's *Motor Trend* in April 1951. The magazine was rather spectacularly failing to pick a motor trend, but it was hardly alone. Indeed a great deal of the twentieth century was spent speculating about a bold automotive future where, by the time the twenty-first century rolled around, we'd be silently spirited along by cars that flew, floated and flashed past at supersonic speeds. Well, here we are . . . and look what we are driving.

Disappointing? Well, yes and no.

For a start we should be pleased that we've made it this far. In the early days of motoring, most speculation about the automotive future involved guessing when the bubble

> 'One day we shall endow chariots with incredible speed without the aid of animals.'
>
> English philosopher, scientist, mathematician and theologian Roger Bacon gives his hot tip of 1279. This and other provocative predictions led to accusations that Bacon was in league with the devil. He was jailed until 1292.

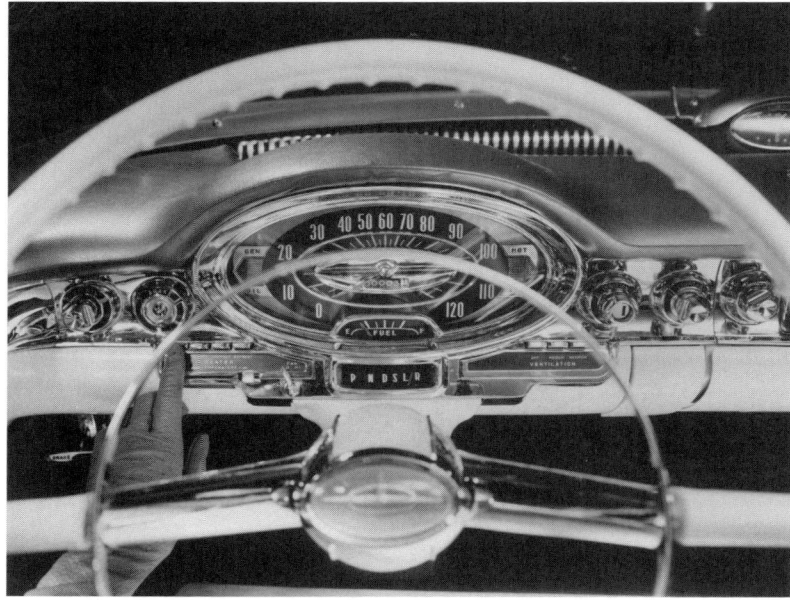

In 1957 Oldsmobile saw the future – and it was very heavily chromed.

'For a very long period before the time of Our Ford . . .'

A typical line from *Brave New World*, the novel by Aldous Huxley (published 1932). In Huxley's scientific utopia, the people worship Henry Ford and are much given to making 'the sign of the T'. The story is set in the year 632 A.F. (After Ford).

would burst. In 1909, *Scientific American* magazine was convinced that 'the automobile has practically reached the limit of its development'. The magazine justified this conclusion by saying no major improvements had been introduced for more than a year.

Fortunately, things did progress: cars became faster, more capable and, in many cases, more outlandish. The Rumpler 'Tropfenauto' (or 'teardrop'), a radical German streamliner of 1921, gave us the first plausible view of what the vehicular future might look like. It also proved from an early stage that almost everything we would ever guess about what was to come would be wrong.

Saint Christopher was the patron saint of travellers and, specifically, 'automobilists' until he was demoted by the Vatican in 1969 (the story of Saint Christopher's life was found to be a bit short on facts). Saint Fiacre, a seventh-century Irish hermit, remains the patron saint of cab drivers.

WELCOME TO THE FUTURE

More breathless speculation: Oldsmobile's Golden Rocket from 1956.

> 'The value of life can be measured by how many times your soul has been deeply stirred.'
>
> Attributed to Soichiro Honda (1906–1991), a man responsible for many cars and motorcycles capable of stirring said soul.

The first 'dream car' produced by a major manufacturer was the 1938, Harley-Earl-designed, Buick 'Y-Job' (probably not a name you'd choose today). It was an oversized, extravagantly styled two-door convertible with hidden headlights and other totally pointless features. But it came at the tail-end of a depression and there was a war on the way. The golden era of 'the future' would have to wait a little longer.

It came with the 1950s and 1960s, spurred along by the lure of a new millennium, the fashion for futurism, and the legacy of a world war that had revealed the possibilities of jets, rockets and nuclear power. There was endless science-fiction-style speculation about what we'd be driving in the year 2000 and beyond – if we weren't slaving in salt

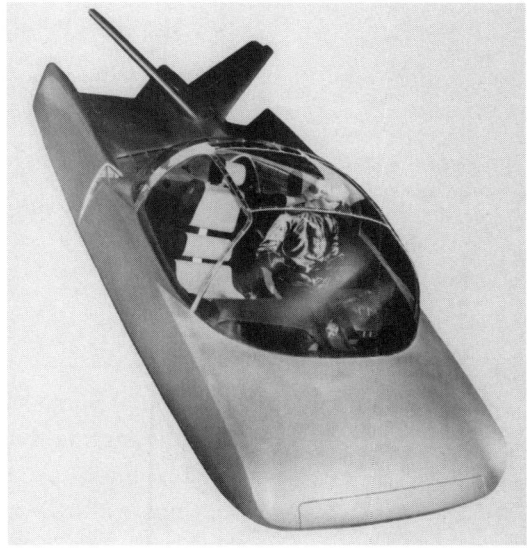

Another prediction of the exciting future, and a European one at that. The Simca Fulgar, supposedly with a range of 5000 km, thanks to six radical batteries. And, yes, there are wheels down there somewhere.

The plan – and these are Ford's words from the early 1970s – was 'elevated guideways as the answer to traffic jams'. On these 'guideways' vehicles would 'move in a steady stream at between 60 and 70 mph under computer control.'

And here is the 21st century reality.

mines for alien invaders.

Artists, designers and futurists took up where the Y-Job left off and produced thousands of drawings, models and even full-size working prototypes. Admittedly, some 'predictions' were just cynical attempts to generate interest at motor shows, however, serious publications such as *Popular Science* were also reviewing with excitement the technology that seemed just a generation or two away: cars that would drive themselves, or hover above obstacles, or use gyroscopic stabilisers, or scoot along raised roads at hundreds of kilometres an hour while serving the passengers hot coffee. And in *Popular Science* and elsewhere there was still a general, unbridled optimism about the car and what it might do for us. It was taken as a given that prosperity would continue increasing, and would bring with it ever higher tailfins.

Alas, within a short time a series of new preoccupations with safety, pollution and congestion brought the age of *more, more, more* to a shuddering halt. We started seeing electric cars, experimental safety vehicles, city minis and

> Several doctors of the late nineteenth century predicted humans could not survive at a speed of 60 mph (about 100 km/h) because neither the human heart nor lungs could cope with the pressure generated. They almost certainly charged for that advice, too.

boring multipurpose vehicles far more often than speed machines with bubble roofs, robotic drive systems and grilles mimicking the air intakes of a jet fighter rockets. Speculation now was about whether the private automobile had finally had its day.

Yet again, though, the car managed to survive and prosper. Sure, today's cars are outwardly less elaborate than what we might have hoped for, but they are better under the skin, more affordable, better handling and easier to drive than was imagined possible during the decades past. We've made many of the primitive wonders of those show cars – such as 'radar cruise control' and 'electric maps' – work properly, thanks to new materials and huge leaps in electronics. We've also managed to produce engines that deliver extraordinary power with fewer nasty consequences, while replacing faux aerodynamics (including those outrageous bumper protrusions and 'powerswept rooflines') with the real stuff, enabling vehicles to cut through the air more efficiently, quietly and with greater stability.

We've also discovered something designers once hardly ever thought about: that having a wonderfully stylish and shiny car is no consolation if you are impaled on a chrome-tipped steering column. Today's cars are almost unbelievably

> 'If the automobile had followed the same development cycle as the computer, a Rolls-Royce would today cost $100, get a million miles per gallon, and explode once a year, killing everyone inside.'
>
> Robert X. Cringley (born Mark Stephens), *Infoworld* (c1997).

Edmund Rumpler's 'Tropfenauto' or 'Teardrop' of 1921 was designed by a plane-maker and looked like a motor launch. It was an early real-world attempt to use the science of aerodynamics.

Chances are that somewhere in another solar system there are cars with digital speedometers, driven by people who think that four-wheel steering is a thing of the future. Or more perplexing still, maybe in a parallel universe, people really do drive the cars envisaged on planet Earth in the 1950s and 1960s. Cars such as the Ford Seattle-ite from 1962.

A few years after pioneering bad car puns with the Seattle-ite, came Ford's Solarus, 'that could be powered by fuel cells energised by the sun'. The company put precisely no effort into making it work. It was too busy selling Mustangs with big and entirely conventional V8 engines.

Peugeot said its Proxima, from 1986, was named for a star in the Centaurus constellation that is 4.27 light years away. Which is roughly how far the Proxima was from production.

good at keeping you out of an accident and protecting you if you happen to be in one. Indeed, if you take subjective things such as styling out of the equation, almost any one of today's cars is better in every department than even the best car in the world was a couple of decades ago.

While all this progress has been taking place, there have been fascinating successes, ghastly failures, brilliant ideas,

And still they try: the 2006 Rinspeed Senso.

monumental misconceptions, spectacular wrong-turns and enough heroes and villains to provide the Brothers Grimm with a sequel. Oh, and some pretty extraordinary motor cars. Which brings us to what *Step On It!* is about. It's a selective, slightly eccentric and hopefully entertaining look at the highs and lows in the career of a machine that has completely changed our world.

It looks at what people have said about cars, thought about cars, and done with cars from 1769 to the present day. As for where things are going, we'll leave that for now. If we've learned anything from the past, it should be that any guesses about the future are almost certain to be wrong.

'Fang² : *verb (t)* 1. To drive (a motor vehicle) at high speed. – *verb* 2. to move (*along, down, etc.*) in a motor vehicle at high speed – *noun* 3. such a drive. [from Juan Fang(io)].'

The definition of a favourite Australian motoring word, from the *Macquarie Dictionary*, Third Edition (1997).

Tens of thousands of patents relate to the automotive industry, a large percentage of them involving alternative engines. The first patent of any type was granted in 1790 for a better way to make pot-ash. In 1899, the Director of the US Patent Office, Charles H. Duell, supposedly declared that 'everything that can be invented has been invented'. In 1980 (or thereabouts) the 4 millionth patent was issued.

CHAPTER 1

Vive la rev

Nicolas-Joseph Cugnot has never become as famous as Henry Ford or Walter Chrysler, but the Frenchman is credited by many – including editors of the *Encyclopaedia Britannica* – as the builder of the world's first working automobile.

Not that the three-wheeled steam-powered beast he created way back in 1769 looks much like anything we'd call an automobile today. It would struggle to generate showroom traffic due to a complete lack of bodywork and an unacceptable dearth of comfort and convenience features. And its performance – top speed just under 5 km/h – could cause all manner of problems with freeway merging.

Cugnot was a military man and when he wasn't writing long and boring treatises on the subject of blowing people up, he tinkered away on various inventions that he hoped would assist in blowing people up. And that was the origin of his 'car', which was really a truck, if it was anything. The Cugnot 'firewagon' was designed to haul artillery.

> **'But when from highmost pitch, with weary car, Like feeble age, he reeleth from the day'**
>
> A rare mention of a car from William Shakespeare. It is from Sonnet VII; the word is an abbreviation of chariot.

Thanks to mounting an enormous steam boiler right at the front of the vehicle, and using steel 'tyres' wrapped around large wooden wheels, Cugnot was also the man who invented understeer. Even at its miserably low top speed his steamer would plough forward with complete disdain for any instructions communicated to it via the twin-handled tiller steering system. In 1771 Cugnot punched a firewagon-sized hole in a brick wall, adding 'first ever motor vehicle accident' to his list of achievements.

Nonetheless, Cugnot had shown that the engine and the wheel could be combined. Others caught on and decided that if a Frenchman could produce something that was expensive, slow, noisy and dangerous, so could they. Britain's Richard Trevithick drove a steam-powered carriage up a Cornwall hill in 1801 and presented his London Steam Carriage in 1803. This was a conventional horse-drawn carriage fitted with a steam engine. To achieve the gearing he needed, Trevithick used 3-metre-tall wheels at the rear, making his machine look like a stagecoach on stilts. It did run on London streets,

Nicolas-Joseph Cugnot's 1769 creation brought us the motor age – and slammed it into a wall.

> Russian revolutionary Vladimir Ilich Lenin gave early credence to the term 'the limo left' by ordering nine Rolls-Royces, including one fitted with a half-track for better on-snow performance.

'Motoring enjoys an enormous vogue in France, principally owing to the absence of police restrictions and to the excellent roads.'

Baedeker's *Guide to Southern France* from (alas) 1907.

however it did so slowly, unreliably and with such expense that it was soon abandoned. Trevithick turned his talents to locomotives instead and somehow ended up in Peru looking for silver before dying penniless in 1833.

Around the world, many more steamers were built with varied success, which is to say success that varied between none and not very much. Then, in 1860, Jean Joseph Etienne Lenoir, a Belgian inventor, produced a two-stroke internal combustion engine. It ran on coal gas and its efficiency was only 4 per cent, but it was revolutionary and it is widely believed Lenoir went so far as to make a primitive automobile in 1862. By one report this completed a 10 km journey over a three-hour period.

Progress was being made and others could see that a lighter, more efficient engine like Lenoir's was a better starting point for a motorised carriage. The Delamare-

The Bouton steam-powered machine of 1860. Neither stylish nor quick, it wasn't reliable or user-friendly either.

In 1984, the French banged up a replica of the Delamare-Deboutteville, a vaguely car-like object that may or may not have been built a hundred years earlier, so it could claim the automobile was a Gallic invention. Anything to stick it up the Germans.

Deboutteville of 1884 is claimed by the French as the first real car, on the basis that it had an internal combustion engine, was light and small by the standards of the massive steamers and that its existence was properly documented.

However, there is a small problem with that claim. The first creation of Edouard Delamare-Deboutteville, a 27-year-old textile producer, was a tricycle supposedly built in France in 1883. But it exploded in the same year, wiping out most traces of its existence. The kaboom was caused by a leak from the container of town gas that was fuelling the engine. The Delamare-Deboutteville car was a 'dog-cart' four-wheeler supposedly produced the next year. By one report it shook itself to pieces on its first run, by another it was never actually built, merely sketched on paper for a patent application. None of this stopped the French building a replica from the patent sketch and holding 'centenary of the

A French inventor, Gustav Lebeau, took out a patent for automotive seat-belts in 1903.

'For most purposes, a man with a machine is better than a man without a machine.'

Henry Ford states the inarguable, 1926.

STEP ON IT!

This is Gottlieb Daimler's 1889 effort. Slow, short on creature comforts and a shocker in the rain, but otherwise the world's first practical four-wheel. The person with the most facial hair was allowed to hold the central tiller.

automobile' celebrations in 1984, a year before the Germans commemorated what they believed was the correct 100th anniversary.

The Germans' claim was harder to argue against. It was based on the work of Karl Benz and Gottlieb Daimler, two engineers working independently and in great secrecy because they were both violating the four-stroke internal combustion engine patent held by fellow German Nikolaus August Otto. Benz completed his three-wheeled car in late 1885; Daimler fitted his engine to a purpose-built two-wheeler and conducted road trials before the same year was out.

The steam car outsold the petrol car in the early days of motoring. This is an 1895 Serpollet. Its noise (or flamboyant styling and apparently random seating configuration) appears to be giving the man standing to the left a heart tremor.

In January 1886, Otto's patent was declared void by a German court and Daimler and Benz could come out of the woodwork, or workshop. Within a few years they were both selling motor vehicles under their own names.

If Benz and Daimler were stern and serious, Karl's wife Bertha provided some light relief. In 1888 she pioneered the illegal use of a motor vehicle, taking a three-wheeled prototype out of the Benz workshop at dawn and – with her two teenage sons – driving it 60 km from Mannheim to Pforzheim. It was a journey not completed until after dark (there were no headlights) and Bertha was too tired to visit her relos in Pforzheim by the time she arrived. However, it ranked as the longest car trip conducted by anyone up to that time.

A viable motor industry was born on the back of the efforts of Daimler and Benz (whose firms merged in 1926 and used the Mercedes-Benz brand), yet it was France that took most strongly to the automobile. Thanks to Renault, de Dion-

This 'pleasure carriage' was 'entirely under the instant control of its driver' and 'capable of any speed within the Government regulations'. In England until 1896 that meant any speed up to and including 4 mph, providing you had a man with a red flag walking in front.

An 1896 Benz in London. Note the briefcases: the new car makers were trying to convince buyers it was a practical business tool and not a rich man's plaything.

Many early models used passengers as protection for the driver in the event of a head-on collision. This is a 1904 Rexette three-wheeler, competing in the annual London to Brighton Rally in 1951.

Racing cars against trains, planes and horses was once common. Bob Burman is shown driving this Buick at Daytona Beach in Florida. It was stripped down but still looked considerably more solid than the plane and not a great deal less airworthy. Note the unusual braking system employed by the pilot.

Bouton, Peugeot and others, there were 6546 cars driving on French roads by 1899. That was ten times as many as in the United States and 15 times as many as in Germany. And it was 6546 times as many as existed in the whole world 130 years earlier when Nicolas-Joseph Cugnot started the whole thing going.

Louis Renault, seen here in the 1901 Paris–Berlin race, used motor sport to prove the merit of his early cars. When his brother Marcel died in the 1903 Paris–Madrid Trial, the company withdrew from motor sport. It eventually returned and more than a century later – in 2005 – a Renault won a World Formula One championship.

Taking sides

Why do the Brits and most of their colonial descendants drive on the left side of the road, while others drive on the right? It is a question that has occupied historians and philosophers for thousands of years. Well, it has certainly bugged a few of us for a while.

One seemingly logical explanation is that carriage drivers traditionally drove on the left side of the road because, being mainly right-handed, this enabled them to swing a whip at the horses without cleaning up a few pedestrians. The first car drivers followed suit.

But then why did the Americans and Continental Europeans start driving on the right side of the road? Some claim that when the steering wheel replaced the tiller, the car maker would put it on either side, depending on the whims of the purchaser. When Henry Ford started mass-producing cars he had to standardise so he arbitrarily opted for the left-hand

> **An 1897 brainchild of one Henry Silvester of St Louis (USA) was a car powered – in theory at least – by air compressed by the movement of the suspension. Was it successful? Er, not as such.**

STEP ON IT!

The car is an 1888 De Dion-Bouton, photographed in Australia. The central tiller and scarcity of other traffic meant choosing which side of the road to drive on wasn't a pressing issue. Child safety restraint legislation hadn't quite been nailed down yet either.

side with the Model T in 1908. Fords quickly became the most popular vehicles on the road. With the wheel on the left, motorists tended to drive on the right to distance themselves from poles and posts at the side of the road.

Another theory also relates to Henry. It is that he deliberately sought to circumvent the so-called Seldon patent. George Seldon – a clever US attorney – managed to secure a patent on the motorised road vehicle in 1895. Although he had never actually built a car, Seldon had been the first to file a 'working drawing' with the US Patents Office. As a result he managed to collect royalties from almost every American car maker. Ford refused to pay and

> The Archduke Francis-Ferdinand of Austria-Este, when shot by Gavrilo Princip in Sarajevo in 1914 (precipitating World War I), was travelling in an Austrian car known as a Gräf and Stift. It was a 1910 model, rated at 28 hp. An English newspaper of the 1960s offered a prize for the best imaginary headline. The winner: 'Archduke Found Alive: World War I Fought By Mistake'.

Pre-20th-century motoring at Manly, Sydney. Right, left or centre: it was a matter of finding a gap.

eventually defeated Seldon in a long court battle. The theory goes that putting the wheel on the left was one of the many tricks Ford used to make his car as different as possible from the Seldon-licensed cars then being built.

Yet another is that while empire-building in Europe, Napoleon (himself a southpaw) exercised the usual French contempt for everything Anglo by decreeing that all European carriages would drive on the opposite side of the road to English carriages. Though this doesn't explain why in the 1920s many French luxury car makers were still exclusively producing right-hand drive cars, while in Austria half of the country drove on the left, half on the right. This stayed the case until Hitler '*anschlussed*' Austria in 1938 and standardised practices with Germany. Mussolini made all the traffic in Italy run on the right of the road. Until then, it had kept to the right in the countryside but to the left in many cities. In Sweden they continued to drive on the left side of the road in left-hand drive vehicles until 1967.

Dr Maxwell G. Lay, in his highly entertaining book *Ways of The World* (Primavera Press, Sydney, 1993), suggests the first law on the subject was in China in 1100 B.C., dictating that the right side of the road was for men, the left-side for

'Don't travel on the wrong side of roads around curves.'

'Don't get into the habit of looking round behind you – have a mirror.'

'Don't look into the petrol tank to estimate its contents with a lighted cigarette in your mouth.'

Three of the many snippets of sage advice to be found in the book *Don'ts for Motorists*, published in 1924 by W. Foulsham of Fleet Street, London.

Henry Ford wrote about his manufacturing systems in the 1926 *Encyclopaedia Britannica*, in the process coining the term 'mass production'. However, the process had been seen in the clock, sewing machine and gun industries before it was used in building cars. In the late 1980s the term 'mass customisation' was used to describe the ability of modern manufacturers to produce vast numbers of models and variations on the same production line.

> 'The negative attitude of cars is expressed in their very name, automobiles, which exalts the vehicle at the expense of the person transported by it. They are symbols of machismo, aggressivity and empty consumption.'
>
> Noted lemon squeezer designer Philippe Starck (born 1949), as quoted in *Architectural Digest Motoring*. Starck told the magazine he owns a Mini Moke, adding: 'Despite its military origins [the Moke was designed for the British Army] it's never been macho. And it's androgynous – sexually neutral and politically benign. Is it beautiful? No. But I'm not interested in beauty for its own sake.'

women and the centre for carriages. Between that date and the Middle Ages there was so little traffic that in most places it was not necessary to legislate. Lay says the choice of which side cars travelled on was partly determined by what type of carriage was most common in various parts of the world. In the United States, the Conestoga wagon, which had been instrumental in opening up the west, had a brake lever of the left-hand side of the vehicle, requiring the driver to sit on that side. To gauge clearances for passing oncoming traffic, it made sense for the driver to stay on the right side of the road.

British legislation in the eighteenth century enshrined travelling on the left and Napoleon indeed spread the 'right' to France and much of the rest of Europe. According to Lay: 'With the introduction of the steering wheel in 1898 . . . car makers usually copied existing practice and placed the driver on the kerbside. Thus, most American cars produced before 1910 were made with right-side driver seating, although intended for right-side driving.'

A June 1913 edition of the US magazine *Century* quotes R.E. Olds (the founder of Oldsmobile and later Reo) on 'must have' features:

'The leading cars this year have left-side drive. You know that all cars must follow. The delay on some cars is simply due to the cost of changing old-style models. The laws in

Europe compel the driver to sit close to the cars he passes. And he sits there now in the best cars built in America.'

The other 'must-have' features were 'Big Tyres, Set-in Lights, Single Rod Centre Control Gear Lever, Fine Finish'. And surprise, surprise, all the features mentioned were available on the Reo.

In which we swerve

One type of vehicle often overlooked by historians is the dodgem or bumper car, the history of which goes back to at least 1920. Before there could be dodgem cars, however, there needed to be amusement parks. These grew out of the 'pleasure gardens' found on the edges of large European cities in medieval times. Pleasure gardens contained picnic areas and hosted dancing, fireworks, human freaks, music and primitive amusement rides. In the USA in the 1800s, electric trolley bus companies built slightly more sophisticated entertainment venues at the end of the trolley lines. This stimulated weekend travel and profitably used the electricity the trolley companies purchased for a fixed sum per week.

The 1893 Chicago World Fair introduced the Ferris Wheel

All the fun of the fair: the trusty Dodgem.

and a ride-based amusement park opened at New York's Coney Island in 1895. By 1919, over 1500 amusement parks were in operation in the USA and, in 1920, Max and Harold Stoehrer of Methuen, Massachusetts, invented the Dodgem. This large rear-steering vehicle with oversized bumpers was advertised as 'the repeater of all repeating rides . . . it gets them and it holds them'.

The fun that could be had smashing one vehicle into another without guilt, injury or repair bills made the Dodgem an instant hit, and brought many competitors into the market. The most successful were cousins Joseph C. and Robert J. Lusse, whose Auto-Skooter was far more controllable than the Dodgem and, more importantly, better looking. The Auto-Skooter was covered by 11 patents. It was completely surrounded by a large rubber 'bumper rail' but its biggest innovation was that it could be driven away from a collision simply by turning the steering wheel 360 degrees (most other bumper cars required the driver to engage some sort of primitive reverse gear).

When General Motors collaborated on the construction of a roller-car ride based on its crash test facility, it was joining a tradition even older than dodgem cars. The first modern-style roller coaster was built in America in 1884; the first with fierce rises and falls, the Drop-the-Dip, appeared at Brooklyn's Coney Island in 1907.

Variations on a scream: BMW collaborated on this Mini roller coaster based on the 2003 remake of *The Italian Job*.

The archetype was thereby established and the 1930s were boom times for the Lusses and other bumper-car makers. After the war the machines became more car-like, thanks to headlights and new curvaceous fibreglass bodywork.

By the 1970s, however, the US manufacturing industry was on the skids and this applied to dodgems even more than most things. They were perceived as passé in a world full of increasingly complex and exciting 'danger rides'. While America lost interest, Italian makers further developed the concept and built up a market domination they still hold today.

> A Michelin slogan from 1891 proclaimed, 'The air-filled tyre is, and by its nature will always be, faster than other tyres'. Incorrect: the tyres used on the supersonic *Thrust SSC* jet-car were made from solid aluminium. Formula One tyres are filled with a nitrogen-rich dry gas to overcome the inconsistencies caused by the varying humidity levels in that old-tech substance, air. And in 2005 Michelin itself announced an all-metal tyre called 'Tweel'.

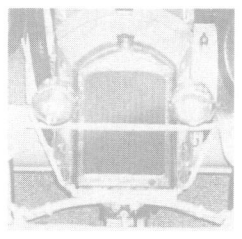

CHAPTER TWO

Turning up the volume

> 'Americans are a broad-minded people. They'll accept the fact that a person can be an alcoholic, a dope fiend, a wife beater, and even a newspaperman, but if a man doesn't drive there's something wrong with him'.
>
> Art Buchwald, Pulitzer Prize-winning newspaper columnist (born New York State, 1925).

The first car produced in substantial volume was America's Oldsmobile Regular Runabout. It was a crude buggy with a one-cylinder engine and tiller steering. It looked worryingly frail and 10 per cent uglier than sin. And it had an oddball 'curved dash' at the front which looked like it had been borrowed from a snow sled.

Yet the curved-dash Olds represented a whole new approach to car manufacturing. Until then, car companies had generally hand-built their vehicles, adapting them to the precise requirements of the buyer. By using a fixed design Oldsmobile slashed the price and expanded the car market beyond the obscenely rich. Now the run-of-the-mill wealthy could afford cars too and, in 1901, 425 examples of the Olds were turned out. In 1902 no less than 2500 were produced and a figure of 4000 was achieved in 1903.

So if it was Germany where the motor industry was born and France where it first boomed (and where Louis Renault had laid some of the foundations for assembly line production), it was in America, thanks to cars such as the Oldsmobile and the evolving art of mass production, where it came of age.

With its 1903 tally, Oldsmobile accounted for 36 per cent of all US petrol car sales with just one model. Mind you, petrol cars weren't the only thing Americans were buying at the start of that century. In 1900 there were nearly twice as many steam cars manufactured in the USA as there were

The curved-dash Oldsmobile of 1901, generally agreed to be the first mass-produced car, developed just over 2 kW from 1.5 litres. Today, a plain-jane hatchback will develop 40 times as much power from the same engine capacity. It will give better weather protection too.

> **Henry Ford may have popularised the saying 'Any colour you like as long as it's black' but the early Model Ts were delivered in 'Brewster Green'. Model T Fords became universally black only after it was discovered that 'Black Japan Enamel' dried quickly and could be applied by spraying. From the mid-1920s, thanks to improvements in paint technology, buyers could again order a standard T in colours other than black.**

The Model T was not much more plush than the Olds, but it went on to be the most successful car of the early era by a huge margin. This is Henry Ford with his first car – the 1896 Quadricycle – and the 10 millionth 'Tin Lizzie'.

'History is more or less bunk. It's tradition.'

Henry Ford (1916). The quote is of course remembered as 'History is bunk'. As a sequel, in 1968 Henry Ford's grandson, Mr II, enticed a top General Motors executive to Ford in controversial circumstances. His name was Semon 'Bunkie' Knudsen and 18 months later Henry junior sacked him in the belief he was setting up a spy system within the company. A graffitist at the Ford plant chalked up: 'Bunkie is History'.

petrol. Even electric cars notched up 1575 sales against petrol's 963.

But it was a short-lived situation. Even the most successful steam car companies – including the 'Stanley Steamers' built by twin brothers Francis and Freelan Stanley – could not produce vehicles to match light, cheap, simple and reliable petrol cars such as the curved-dash Olds. Electric cars would survive a little longer than steamers because they were simple and easy to drive (no clutch or gear lever required) and because, unlike petrol cars, they didn't need to be hand-cranked. Cranking could be a dangerous pursuit for a strong man, and was certainly considered no job for the sort of wealthy matron likely to be an early female motorist. Even if you didn't break an arm, you could easily split a nail.

Explaining what happened next is a little like painting the battle of Waterloo on a postage stamp, yet it is still possible to plot the significant points, most of which bring you back

In 1916 the USA became the first country in the world to build 1 million cars during a single year. The second country to achieve the feat was Britain in 1954.

Electric cars – this is a 1916 Rauch and Land, built in Ohio – were advertised as being ideal for female motorists because they didn't need to be cranked by hand. The formidable woman in this publicity shot, however, looks like she could hand-crank the Queen Mary.

Charles F. Kettering (pictured centre) brought sales of battery-powered cars to a near halt by inventing the electric starter. Kettering also came up with an aerial torpedo, a treatment for venereal disease and ozone depleting Freon (CFC) gas.

The Chevrolet company was founded by Swiss racing driver Louis Chevrolet but it was his younger brother, Gaston, who achieved the biggest victory on the track. Gaston won the 1920 Indianapolis 500 in a Monroe-Frontenac.

This brave early attempt by Peugeot to reduce frontal area had only two drawbacks: the driver had to tuck his elbows under his ribs and the sole passenger had to sit in the back. Still, there are not many cars you can park in the hallway when the garage is full. For the record, it's a Peugeot Type 161 from 1921.

to one odd but brilliant man: Henry Ford. In 1904, the year after Henry formed the Ford Motor Company, US production of motor vehicles overtook that of France. By 1913, thanks largely to Ford, no less than 80 per cent of all the cars built each year were made in the USA.

Thousands of other car companies around the world rose and fell before World War I, but none had Ford's influence. Henry Ford, and the many clever but little-acknowledged Ford company engineers and marketers who didn't get their names on the badge, set out to produce an affordable car that was light and simple, yet made from high quality materials. First there was the Model N then, from 1908, the Model T. By 1913, the moving assembly line was up and running, while machine tools were becoming more complex and specialised in the endless pursuit of higher production.

A slightly more stylish 1939 model from Peugeot adopted the aerodynamic look pioneered – and quickly discarded – by Chrysler in the States.

Herbert Austin, here with his 1922 Austin 7, managed to sell this car as a full four-seater. He was a very clever man.

The Model T was tough, easy to drive, cheap to maintain and, most of all, affordable. The four-seater version was just US$825 when it was launched (the much cruder Oldsmobile with just two seats and a one-cylinder engine had cost $650) and there was better news to come. Economies of scale and Henry's eccentric policy of regularly cutting the price even when he had waiting lists meant that eight years later the T cost just $345, almost 60 per cent cheaper than it was on day one.

From 1912, the self-starter (seen first on Cadillac, initially rejected by Ford, but almost universal by 1920) provided a near-lethal blow to the electric car industry. It didn't help, of course, that the average electric car now cost three times as much as a Model T Ford.

By 1912, the price of a T had dipped below the average annual wage in the USA. In 1915 the millionth Model T left the factory and by 1916 one in every two cars built in the US

> 'The never ending procession of motor traffic is so great that motoring is no longer a pleasure.'
>
> *The Sydney Morning Herald* – in 1929.

was a 'T'. In 1920 the USA finally had as many engine-powered vehicles as it did horse-drawn. Motoring for the masses had truly begun.

In the early 1920s a complete T Ford could be produced from raw materials in four days, but General Motors' Chevrolet division was soon equalling Ford's productivity, and doing it with more modern designs.

General Motors was the brainchild of William Crapo Durant, a flamboyant part-genius, part-nutter who was fascinated by enormous deals and had little interest in the smaller details. Durant obsessively bought up every car maker he could and very nearly grabbed hold of Ford for just US$8 million shortly before the success of the T Model went stratospheric. But after making many bad acquisitions soon afterwards, Durant lost control of GM. He then founded Chevrolet (with Swiss-born racing driver Louis Chevrolet) and

If a V8 is good, a V16 must be twice as good. Cadillac introduced its V16 engine in time for the Depression, yet managed to flog a few before the crunch really hit. It survived in various model series until 1940, but sales by then were just a trickle.

> The best known Reo cars had names like Flying Cloud and Wolverine but the company also built the Reo Speedwagon, a truck that would have been forgotten if not for a rock band that should be.

used this as a vehicle, so to speak, with which to regain management control of General Motors.

The GM group included Buick, Cadillac, Oakland (which would become Pontiac) and that pioneer of US volume production Oldsmobile, which GM acquired in 1908. By that stage, however, founder Ransom Eli Olds had left to form a new company, which was named after his initials: Reo. Within a few years Reo was one of the four biggest brands in the US (alongside Ford, Buick and Maxwell-Briscoe).

By the 1920s 'Billy' Durant had imploded again and General Motors was being run by Alfred P. Sloan Jr. With the 'Under new management' sign hung on the door, Durant's sprawling, haphazard concern was quickly evolving. Sloan laid the foundations of the modern corporate structure and created the graduated model line-up. By having a range of brands and variants within each, Sloan was aiming to provide 'a car for every need and every pocketbook' or, to put it in plainer terms, to gouge the maximum fee from every potential buyer in every potential price class.

After World War II, the influence of the USA on international motoring diminished. American cars had become too large, heavy, thirsty and expensive for the rest of the world. For car ownership to move to the popular sphere in Europe and elsewhere after the devastation of the war years, a new type of light, fuel-efficient vehicle was needed. Europe and, later, Japan were going to provide it.

> **'It was a rich cream colour, bright with nickel, swollen here and there in its monstrous length with triumphant hat-boxes and supper-boxes and tool-boxes, and terraced with a labyrinth of windshields that mirrored a dozen suns.'**
>
> F. Scott Fitzgerald describes an automobile belonging to his best-known character, Jay Gatsby, in the classic 1925 American novel *The Great Gatsby*. The brand of car was not specified.

Eccentric titans

Born in 1863, **Henry Ford** made his money from industry but spent a lot of it glorifying agriculture (he even constructed Greenfield Village, a town idolising the time when the USA was almost solely agrarian). In 1915 he chartered a ship to take him and other pacifists to Europe, where they believed they could end the war by 'continuous meditation'. Ford came within 2200 votes of becoming a US Senator in 1918, despite a refusal to campaign on his own behalf (he was beaten by Truman H. Newberry); soon after, he was attacking Jews, blacks and immigrants in his own ironically named newspaper, *The Dearborn Independent*. Ford dramatically increased wages for his workers and reduced their hours, yet met unionisation with violence. He cut the price of his cars even when it made no business sense, because he genuinely wanted everyone to enjoy the benefits of the motor age.

Henry Ford at the wheel of his six-cylinder racer in 1905.

TURNING UP THE VOLUME

> The Cadillac company was named after the man who founded Detroit: Le Sieur Antoine Laumet de la Mothe Cadillac, Knight of the Royal & Military Order of St Louis. There is some doubt, however, as to whether the grand title was real, or whether 'Cadillac' was a humbly born adventurer who had upped his station in the New World.

Although Ford had lost many of his marbles in his latter years, he was dragged back to lead the company in 1943, following the early death of Edsel Ford, his son and replacement as Ford President. In 1945, grandson Henry II turned 25 and took up the reins; two years later, the old man ceased production, aged 83. When Henry Snr died, his room was lit by candles to save money. He had US$26.5 million in the bank.

Originally a carriage manufacturer, **William Crapo Durant** was the owner of the strangest middle name in transport history. He made and lost several fortunes. His success at the struggling Buick led him to form General Motors in 1908, taking on component manufacturers as well as car makers large and small. Durant received the boot from GM in 1910, then regained the tiller in 1915 having co-founded the Chevrolet brand in the interim. In 1920 Durant once again became an ex-GM employee and was finally and thoroughly cleaned out in the Wall Street Crash of 1929. In his final years, Durant was energetically trying to make a buck out of 'Bowl-a-rama Billy' franchised bowling alleys. His Chevrolet co-founder, Swiss-born racing driver Louis Chevrolet, had an even steeper fall from grace: he ended his working days as an assembly line mechanic for the company that bore his name.

The man who formed General Motors, William Durant.

Louis Chevrolet and his 1911 car known as – stop me if you've already guessed – the Chevrolet.

Walter Percy Chrysler was a one-time railroad machine shop apprentice who rose to be boss at Buick, by then part of General Motors. When Chrysler wanted to quit in 1916, GM boss William Durant upped his annual salary from an already extraordinary $50,000 to an almost unbelievable $500,000. Chrysler was paid even more at rival car-maker Willys soon

Walter P. Chrysler.

after – $1 million a year – and showed his appreciation by slashing the salary of founder John Willys to $75,000 and trying to engineer a takeover. Chrysler formed his own company in the mid-1920s from the hulk of the ailing Maxwell concern. He also commissioned New York's Chrysler Building, which was the tallest building in the world when completed in 1930, and the only one inspired by the automobile. It has stainless steel gargoyles styled like radiator mascots, plus hubcap patterns and other motoring motifs worked into the exterior design.

Louis Renault, the son of a draper-cum-button manufacturer, was obsessed with everything mechanical and built his first car in 1898. His litany of eccentricities included a refusal to borrow money from any source. Instead of using the usual network of supplier companies, he built all

Mercedes Jellinek gave her first name to the car. Which is lucky, because a Jellinek-Benz doesn't have much of a ring to it.

Louis Renault at the wheel of his 1898 Renault Voiturette. Or, perhaps, an exact small-scale replica of his 1898 Renault Voiturette.

> 'We are the first nation in the world to go to the poorhouse in an automobile.'
>
> Will Rogers, American actor and humorist (1879-1935). The comment referred to the USA and the Great Depression.

the parts for his cars – even the wrapping paper for the spare parts and the bricks for his factories.

The founders of Volvo, **Gustaf Larson** and **Assar Gabrielsson**, met in 1924 at a crayfish party (as Swedes do) and decided to build a simple, four-cylinder car specifically for Nordic conditions. Rather than opt for the huge suites of leather and mahogany-lined offices popular among US executives, the pair shared the same single, austere desk until they retired from day-to-day operations three decades later.

Early achievers

1899: Henry Sutton of Melbourne, Australia, builds a car with front-wheel drive. Production runs to, approximately speaking, one.

1900: Fiat develops all the elements of a modern power steering system.

1903: De Dion-Bouton fits a gearbox employing the basic principles of the modern automatic transmission.

1905: Rolls-Royce sells cars with a V8 engine – and a station wagon.

1907: Sydney's Alfred John Swinnerton builds a car with the chassis frame and body integrated into a single unit to add strength and save weight.

1909: Sizaire-Naudin produces independent front suspension.

1914: Delage builds a double overhead camshaft engine with four valves per cylinder.

1922: Lancia produces a monocoque body and integrated boot.

1933: Mercedes sells the Ferdinand Porsche-designed Model 130 family car with rear engine and many features later used in the VW Beetle.

1934: Citroën unveils the Traction Avant, the first mainstream front-drive family car. The cost of its development (and a gambling problem) sent company founder Andre Citroën bankrupt and he died soon after, believing it was a failure. It went on to sell 750,000 examples over 23 years.

> If those who ignore the lessons of history are bound to repeat it, then BMW can rest easy. No other car company is as scrupulously honest about its past. The German company's official history, *The BMW Story*, goes so far as to show Dachau concentration camp inmates being forced to work on the production line at BMW's Allach plant. A photo from 1936 shows almost the entire staff of BMW's Munich plant giving a Nazi salute.

Remembrance of things fast

When more than one car existed, it seemed only logical to match one against the other, to prove once and for all that mine was faster than yours.

It is generally agreed the first motor sport event was a so-called reliability run, conducted from Paris to Rouen in 1894. The distance was 80 km, and the winner achieved an average speed of just over 17 km/h.

France also hosted the first event formally described as a race, held from Paris to Bordeaux and back in 1895. The distance was 1178 km, and average speeds escalated to above 24 km/h. Okay, 24 km/h was a fair bit shy of the speed of light, but to maintain such an average in the days of mechanical uncertainty, solid tyres (or constant punctures), and unmade roads, you had to take serious risks. Soon all manner of car, tyre and oil companies were holding long-distance speed trials and reliability runs to prove that the

> **'Streets full of water, please advise.'**
>
> American writer Robert Benchley telegrams his editor. Benchley (1889-1945) had just arrived in Venice.

There was a regulation in the 1909 Prince Henry Trial – it seems – that you had to wear goggles, but couldn't use them to cover your eyes. One man in the car following this NSU is clearly cheating.

> 'You will drive even faster on the highways of heaven.'
>
> Tombstone of Italian racing driver Tazio Giorgio Nuvolari (1892–1953).

horseless carriage was a worthy substitute for the horse and carriage. Standard and heavily modified motor vehicles also raced horses, trains, planes and the clock.

Most of the early events were on public roads, causing a large number of fatalities among competitors, spectators, livestock and uninvolved people trying to cart their grain to market. Authorities around the world started stamping down on such events in the early years of the twentieth century, moving much of the action to closed circuits.

The world's first purpose-built sealed motor circuit was the Brooklands track, constructed on private land at Weybridge, England, in 1906–07. Lap speeds achieved on this steeply banked track exceeded 160 km/h well before World War I. However, in his book *Ten Years of Motors & Motor Racing*, pioneer race-driver Charles Jarrott ventured that what he called 'the greatest sport evolved by man' was already in terminal decline:

'The degeneracy of motor-racing as a sport is due to the financial issues now involved in each race – the immense value of victory and the commercial disaster of defeat . . . I can see in the near future . . . the sporting element obliterated altogether by the all-devouring monster of commercialism.'

When Jarrott wrote these words, so similar to the gripes we hear about all sports today, it was still only 1906.

In 1928 Fritz von Opel reached 238 km/h at Berlin's Avus racetrack in the experimental Opel RAK, the world's first rocket-powered car. Later, RAK3 ran on rails and pushed the speed up to 254 km/h. In 1929 GM Corporation bought the Opel company and ended the rocket car program. Spoilsports!

STEP ON IT!

Early racing, Stateside. This is Barney Oldfield at the wheel of a massive '200 horsepower', 21.5-litre Blitzen Benz at Daytona Beach circuit in 1910.

In Europe at the same time they tried to keep engines smaller by reducing the size of the allowable bore. The solution? Longer strokes, as shown in this ridiculously tall-bonneted LION-Peugeot. It hit 152 km/h and might have gone even quicker if the driver could see where he was going.

> America's most famous circuit, the oval at Indianapolis, opened in 1909. It was paved with bricks in time for the first 'Indy 500' in 1911 and quickly became known as 'The Brickyard'. When an oval racetrack was built at Walt Disney World (Florida, USA) in the 1990s, smart types christened it 'The Mickyard'.

CHAPTER THREE

The drive-in culture

It was a remarkably short hop from the horseless carriage to the roadside hamburger. And the culinary world wasn't the only one to be transformed by the car.

The horseless carriage – that rich-man's plaything, that noisy fad – would go on to confound those who dismissed it as a passing fancy, and would have a dramatic effect on leisure, entertainment and even architecture.

From the early years of the twentieth century, buildings were increasingly shaped to appeal to motorists speeding along the road rather than the person on foot or ambling along by carriage. Roadside businesses, restaurants and service stations developed into three-dimensional billboards, identifiable from hundreds of metres away.

Houses changed too, the gardens revamped to include a motor house (or 'garage') and a driveway. It could even be argued that the old social order was been broken down. You no longer needed to make friends out of your neighbours, which for

> 'The Red Guards even changed the traffic lights, so that red became the symbol for "go", until Zhou Enlai told them that red got people's attention better so should remain "stop".'
>
> Author Philip Short on the madness of China's Cultural Revolution in *Mao, A Life* (John Murray Publishers, 1999).

STEP ON IT!

> **La Triviata:** The Alfa Romeo badge depicts a man being swallowed by a snake. The symbol came from the coat of arms of the city of Milan, the company's headquarters.

many people was a great relief. Just as conveniently, there was now no need to stick to your own neighbourhood for entertainment or leisure activities. You could easily haul the family around in the car – or start a new one (one oft-quoted, but hard to source, statistic suggests that one in four conceptions takes place in a motor vehicle).

No longer did you need to buy locally. Combined with another major machine of the twentieth century, the refrigerator, the car revolutionised shopping patterns. People could buy a boot-load once a week, rather than grabbing fresh food every day. This became even easier when there was just one location in which to do it, rather than a succession of specialty shops. Enter the Piggly Wiggly, the

No, we don't know why either. But this odd man modified his Austin 7 to resemble a boat and drove it through the London traffic each day.

first of which was set up in Memphis, Tennessee, in 1916. The founder, Clarence Saunders, patented the idea, which involved the customer passing through a turnstile, following a route between shelves, selecting his or her own items and paying for them all in one hit at a 'checking counter' at the end.

By 1930 there were over 2600 Piggly Wigglys in the USA, plus plenty of imitators. Why was it called Piggly Wiggly? 'So people would ask that question,' Saunders once quipped. Meanwhile, in 1933 the word that would become synonymous with the concept was coined by Albers Super Markets in Cincinnati. The early 1930s also brought the first drive-in

Londoner Louis Mattar was another early car customiser, making so many changes to his 1947 Cadillac that he could talk on the phone while smoking a hookah. Note the whisky dispenser and reel-to-reel tape player. The television installation, you'll have to agree, is super-neat, and note all the extra buttons and dials across the top of the dash preventing any ugly views of other traffic.

> Australia was at the forefront of the trend towards putting people on motorised wheels. By 1927 there were more cars per head of population in Australia than any country other than the USA, Canada and New Zealand.

The car brought new freedoms. You could, for example, put your dress on the roof and lie backwards on a ridge of sheet-metal while balancing a floatation device on your foot and holding an umbrella sideways.

cinema, in Camden, New Jersey. The first drive-in freeloaders no doubt hid in the boot of a car a very short time later.

Faith in the special powers of the car was often solidly misplaced. In 1925, a special motoring edition of popular Australian magazine *The Home* predicted that 'the people with cars will relieve the city areas of housing congestion, and benefit themselves by the healthful atmosphere of country life.' We're still waiting.

In the USA restaurateurs adapted to the car with a new type of quick-service roadside restaurant. In 1948 the McDonald brothers, Maurice and Richard, revamped their diner, cut back the menu and started adding assembly line techniques to food production. Hamburgers were built with the same technology as Fords and, cynics may suggest, with much the same taste. By the 1950s the McDonald's name and concept was franchised across the USA. Pizza

With freedom comes responsibility, and new forms of taxation.

THE DRIVE-IN CULTURE

Accidentally dangerous: the first pedestrian death caused by a motor vehicle occurred in 1896, when 45-year-old Bridget Driscoll was stuck by a car that was travelling at about 6 km/h near Crystal Palace in London. The first automobile driver was killed in 1898. According to the Federation Internationale de l'Automobile (FIA), the car went on to claim 30 million lives in the 20th century.

Hut, Burger King, Kentucky Fried Chicken and Taco Bell were among the thousands of companies following in its wake. Soon you wouldn't even need to leave your machine to be fed: just park outside the restaurant and a female 'carhop' – dressed in a fancy uniform and perhaps even on roller-skates – would bring a tray straight to your car window.

In Australia it was the service station chains that propagated 'fast-food'. From 1957 Golden Fleece started building up a collection of roadside restaurants that numbered 140 by 1975. The American-style restaurant chains arrived down-under in the 1970s with considerable success and by the 1980s they had wiped out (or replaced) most of the restaurants at service stations. However, fuel outlets would become host to another phenomenon: the 24-hour supermarket.

But nobody has taken to the

Car customising was another expression of the new car culture. Model T and Model A Fords resurfaced as 'hot rods' during the 1950s and 1960s.

> The Lancia Aurelia B20 GT of 1951 was the first road car to use the name 'GT'. It stands for *Gran Turismo* or 'Grand Touring'.

> '*We were convinced we had to educate US customers not to drink coffee while they drove. They responded by buying the other guys' cars.*'
>
> Mercedes-Benz's Jurgen Hubbert explains why all US-bound Mercs are now fitted with cup holders (1998).

drive-in (or 'drive thru') concept quite the way the Americans have. The land of the free boasts drive-in churches, wedding chapels, banks, and – pioneered for the 2004 US presidential election – drive-in voting booths. On the other hand, the once phenomenally successful drive-in cinemas have almost faded away. The convenience of video cassette and DVD players, cable television, and the effect of rising land prices have combined to bring the metaphorical curtain down.

Another thing motorisation changed forever was holidays. For a whole family to travel by rail has always been expensive and in the early days of motoring, hotels away from the main tourist centres were equally prohibitive for the average worker.

With the affordable Ford Model T, holiday makers were no longer slaves to trains and hotels. They could choose exactly

Jay Ohrberg at California Show Cars produced dozens of customised cars, including the Bat Cycle and hot rods built around baths and bunk beds. This was a more modest proposal known as the Sand Draggin.

The scariest word in art: an installation. This piece of car culture was outside Notre Dame cathedral in Paris, 2000.

when and where they went and, as a result, cars made holidays the preserve of a much larger percentage of the population. The new breed of 'motor gypsies' (the term was a common one) could choose their place and their pace, and take four, five or more people at not much extra cost.

Motor gypsies cooked by the roadside and camped out in

> '**Motorists are the victim of a good deal of unjust obloquy, because of the conduct of a few "hogs", who should not be allowed to drive a wheelbarrow.'**
>
> Melbourne's *The Age* newspaper (1915). It was arguably the first use of the term hog for someone who blocks traffic, and hopefully the last use of the word obloquy.

An early camper, built in Texas on a 1910 Ford T chassis. All that's missing is a porch on which to play the banjo.

What Australians call a speed hump, the Kiwis call a judder bar and the French a *dos d'âne*, literally 'the back of a donkey'.

Caravanning in the 1970s was enhanced by denim tops with wide lapels and flares. Shoes, though, were optional.

tents. Soon entrepreneurs in Australia and elsewhere saw the chance to cater for this new market with slightly more luxurious accommodation right next to the road. At first this involved little more than pegged-out campsites with running water and shared stoves. After a while, permanent sleeping cabins started to appear; by the 1920s these developed into the first motels.

Cheaper than hotels, but more luxurious than cabins, motels spread right across the USA in the 1930s. Best Western was an early success; Holiday Inn followed. The anonymity of the motel – well outside town, no real name required – made them the ideal place for illegal business dealings and for 'Mr and Mrs John Smith' to entertain themselves. Legendary gangsters Bonnie and Clyde used them frequently (and, when travelling, chose to drive Ford cars, according to a famous letter supposedly written by Clyde Barrow to Henry Ford). The reputation of motels became such that in a 1940

And the caravans of the 1970s boom era were big too.

THE DRIVE-IN CULTURE

The USA's *African Americans On Wheels* **magazine is described as a 'national automotive publication for African-American car buyers'. Some may question the rationale behind such a magazine in the twenty-first century, but until about 1960 a publication called** *The Negro Motorist* **listed what America's black motorists desperately needed to know: which service stations, restaurants and motels would serve them.**

It is widely assumed that tyres have always been black. In fact, some of the earliest examples were white. In the late 1950s or early 1960s that some bright spark first marketed luminous tyres. The idea generated an enormous lack of interest in its day and has continued to do so ever since. Still, it looked good in long exposure photography and made it possible to read a street directory at night by the light of a shining Goodyear.

magazine article FBI boss Edgar Hoover denounced them as 'Camps of Crime'.

Another increasingly common sight on the roads were 'motor caravans'. One of the first seen down-under was a three-berth model made in the 1920s for the Bailey family of Deloraine, near Launceston. It was produced by a local boat-builder, who used Tasmanian oak, timber and painted canvas, plus an axle and wheels from a Model T. After World War I, caravans and caravanning became a huge industry. A further revolution came with the campervan, often built around the VW Kombi and filled with

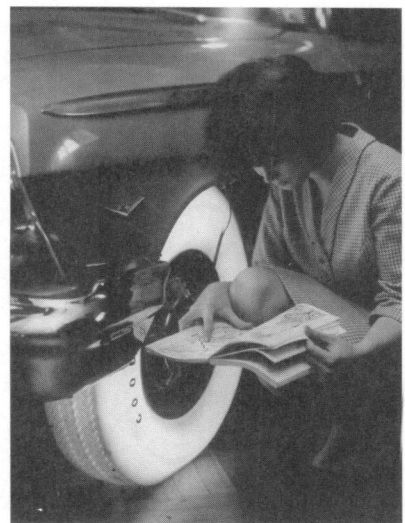

STEP ON IT!

people who thought loose-fitting tie-dyed clothing was attractive.

The 1950s and 1960s were boom years and in 1976 a record 38,000 caravans were sold in Australia. A dramatic crash followed, but since the mid-1990s, the industry has been growing strongly each year. It is estimated that around 350,000 caravans and campervans are now in use in Australia. Most of them can be seen travelling very slowly in the overtaking lane.

Route 66

Route 66, one of the most famous roads in the world, was established in 1926 and linked America's Midwest with the Pacific Coast, working its way through eight states and three time zones along the way.

In the 1930s Route 66 was the escape route out of the

Chuck Berry: motor troubadour.

THE DRIVE-IN CULTURE

> **Early adopters:** Wind-up windows (1919, Pierce Arrow and Packard models); 'Mechanical refrigeration', or air-conditioning (1939, Packard); Self-cancelling indicators (1933, Singer); Electric windows (1948, Cadillac).

Dustbowl; in the 1940s it carried hundreds of thousands of American soldiers en route to the Pacific War. In the 1950s families packed into their Chevy and Dodge station wagons and headed along Route 66 to the newly opened Disneyland complex in California.

A great deal of America's car culture grew up alongside this famous thoroughfare. It was there that the earliest and most ambitious roadside diners, drive-in cinemas and service stations were found. There was even a song to sing along the way, one about a road that wound '2000 miles' from Chicago to LA. Listeners were urged get hip and get their kicks on Route 66. The words and music were written in 1946 by musician and actor Bobby Troup, who claimed to have bought a 1941 Buick with his first song-writing royalty cheque. He used it to head off to Los Angeles, partially writing 'Route 66' in the car along the way. The song was recorded by Nat King Cole, among others, before Chuck Berry rocked it up to produce perhaps the best known version.

By the time Berry's recording increased affection for Route 66, it was 1961 and the road was already in decline. The huge increase in traffic that occurred during the 1950s inspired the construction of a new series of interstate highways to supplant a route that was still mainly two lanes. By 1985 Route 66 was completely bypassed and now only scattered sections remain. In the year 2000, the US Congress passed a preservation law and allocated US$10 million to restore many of the landmarks along the way.

> **'A man who, beyond the age of 26, finds himself on a bus can count himself as a failure.'**
>
> British Prime Minister Margaret Thatcher does her bit for the car industry during a debate in Britain's House of Commons (1986).

The original Chitty

Yes, there was a film: Ian 'James Bond' Fleming wrote the book on which it was based, Roald 'Charlie and The Chocolate Factory' Dahl wrote the screenplay, while Dick Van Dyke and Robert Helpmann starred as goodie and baddie. But decades earlier there was a real life Chitty Chitty Bang Bang. It could neither fly nor float but, that aside, it wasn't much less fascinating than the magical one depicted in the 1968 film.

The real Chitty came from the time after World War I when thousands of unwanted RAF and European aviation engines were offered for sale at bargain prices. Many ended up in preposterously proportioned cars, some of which were used for land speed record (LSR) attempts on public roads not even closed to traffic. Others powered vehicles that that turned up at Brooklands and other early race circuits and achieved ludicrously high speeds, or bit back and killed

Chitty Chitty Bang Bang. This is the film car, built by Ford and pictured with Australian race driver Frank Gardner, who was involved in its creation, and a Ford Escort, which just wanted to get its nose in the shot. Chitty's long bonnet hides a short and modern (for 1968) V6.

drivers trying to feed too many kiloWatts and too many Newton-metres through canvas-belted tyres.

Which brings us to the original Chitty, from the early 1920s. It was a chain-driven racing car built in England by the ultra-wealthy Count Louis Zborowski and fitted with a monumental 23-litre Maybach Zeppelin engine. The car's name, supposedly derived from the distinctive exhaust note, was also applied to two later specials built by the English-born, Polish–American Count.

The Count and his engineer, Captain Clive Gallop, were early champions of the maxim that 'there's no substitute for cubic inches . . . except cubic feet'. Their most impressive effort in this regard was not a Chitty, but the Higham Special, which was powered by a 27-litre V12 Liberty aero engine. This deafening monster, modified and renamed Babs, was used by J. G. Parry-Thomas to set a new land speed record of 275.3 km/h (1926), and to kill himself (1927).

Count Zborowski drove his Chitty Chittys from his palatial home near Canterbury to each race circuit with the passenger seats occupied by a retinue of mechanics dressed identically in black shirts and large tartan caps. The first Chitty was fitted with a 'duck's back body'. Armed with the volumetric equivalent of 14.4 hatchbacks under the never-ending bonnet, Zborowski lapped Britain's famous Brooklands circuit

> **'My cars are made to go, not stop.'**
>
> Attributed to Ettore Arco Isidoro Bugatti (1881-1947), responding to criticism of the braking performance of Bugatti cars.

In September 1927 Isadora Duncan made one of the most famous automotive exits in history when her scarf became entangled in the rear wheel of her Amilcar Grand Sport. Cars were not kind to the American-born dancer. After a family luncheon in Paris 14 years earlier, she had asked her chauffeur to drive her children back to their home nearby. The car stalled on a gradient and when the chauffeur climbed out to restart it, the car rolled backwards into the Seine. Duncan's seven-year-old daughter and five-year-old son were drowned, along with their nanny.

> 'Right! That's it! You've tried it on just once too often! Right! Well, don't say I haven't warned you! I've laid it on the line to you time and time again! Right! Well . . . this is it! I'm going to give you a damn good thrashing!'
>
> Television's hotel manager from hell, Basil Fawlty, having failed to persuade his car to proceed, decides to administer corporal punishment (*Fawlty Towers* episode 'Gourmet Night', 1975). Basil was played by John Cleese; the role of his unreliable conveyance went to a Morris delivery van.

at an average of 113.45 miles per hour (about 183 km/h) in 1922. He also managed to remove most of a time-keeper's hand when he backed the car at high speed into an official's box at the edge of the banked track.

Versions two and three were also built on pre-World War I Mercedes chassis, but used smaller engines (a modest 18 litres, then a meagre 14.7). The third actually made use of things other than brute force, including shaft rather than chain drive.

As well as fanging around in his aero-engined muscle machines, Count Zborowski competed on behalf of makers of more conventional cars, such as Bugatti and Mercedes. His father, Eliot, had also been a racing driver with a penchant for competing while dressed to the nines. He died in 1903 when one of his cufflinks became snagged on the hand-throttle of his 60 HP Mercedes at La Turbie hillclimb in France.

In keeping with family tradition (and the general motor sport practice of the day) Louis also chose to compete until he had a fatal accident, which rarely took long in the days of wooden frames and leather helmets. The Count stopped counting in 1924 when he lost control of a factory-entered Mercedes at Monza. He was not yet 30 and, at the time of his accident, was reputedly wearing the cufflinks salvaged from his father's wreck at La Turbie. Until the 1968 film, no further Chitty Chittys went Bang Bang.

CHAPTER FOUR

Australia joins in

From the earliest days of the motorised carriage, a vast number of would-be car builders in countries around the world desperately wanted to launch their own interpretation of the ideal car. Which, of course, would have their name on the radiator.

The Australian story started in the 1890s with such eponymous efforts as the Shearer, McIntosh, Highland, Sutton, Ziegler and Thomson. The last – a steamer built by Herbert Thomson in Armadale, Victoria – was the least unsuccessful. The grand production total was 12.

One man who shouldn't have stuck his name on the bonnet of anything was Isaac Phizackerley, but he couldn't resist. Not surprisingly, the Phizackerley fizzled fast. Harley Tarrant produced classy cars in Australia before World War I, but they were too expensive to sell well. Mr Tarrant instead turned to selling Model T Fords, finding it a lot easier and a huge deal more profitable. Indeed, the Model T provided the

> 'Good luck to the Holden car, and good luck to all those that ride therein.'
>
> Prime Minister Ben Chifley at the launch of the original Holden (29 November 1948).

Brunswick, Victoria, Australia, circa 1905. The noteworthy car is the one in the rear, entirely constructed by its owner, one Mr Harold Green. So few people build their own cars these days. We've gone soft, buying them at shops.

major problem for anyone wanting to produce unique cars for Australia: it was already almost perfect for Aussie conditions, with its durability, simplicity and high ground clearance. And it became cheaper, not dearer, with each passing year. About 340,000 were eventually sold down-under.

Still, scores of Australian ventures took up the seemingly impossible challenge of competing with Henry's Best. Patriotism was the order of the day with many. Consider the

Go on, complain about the state of the roads today. This is one of Sydney's main thoroughfares, Parramatta Road (at Ashfield), photographed in 1920.

Australis, the Australia, the Pioneer and the Roo, each with a name a little dafter than the previous. The first example of the Roo was launched with great fanfare in 1917, a second example was built with considerably less, and a third was half-completed when the operation fell over.

Most other early car-builders also ran out of steam after building one or two examples but some managed to produce larger totals and thereby increase the size of their financial loss.

The usual trick was to buy the mechanical components from overseas, then mate them with local chassis components and bodywork. This was the technique used with

The Model T could be easily adapted to all sorts of things. This Tasmanian farmer made a tractor from his.

Early motoring could involve crossing rivers, animals and even small children.

Showjumping with a difference. The man under that rakish hat is Australian record breaker (and suspension breaker) Norman 'Wizard' Smith.

the Lincoln, Summit and Australian Six of the early 1920s. All were entirely creditable efforts but doomed from the outset. Ford and GM set up assembly operations in the 1920s putting local bodies on imported chassis at prices no local maker could hope to match. The coming of the Holden, built by the General Motors-Holden's operation from 1948, finally proved it was possible to profitably manufacture a unique car in Australia, as long as you were backed by the resources of a huge American conglomerate and possessed sophisticated facilities built up during war production. There were still plenty of people who missed the message. Shockers of the 1940s and 1950s included the two-stroke, 3.5kW JB Minor, the three-wheeler Tilbrooks and the ridiculously unstable Edith minicar, which seemed to have been modelled on a roll-top desk.

Italian-American driver Ralph de Palma and his riding mechanic, Australia's Rupert Jeffkins, push their Mercedes across the line at the 1912 Indianapolis 500. Later rules allowed the cars to be driven. OK, that's not entirely true: the pair had led for 196 of the 200 laps but suffered a broken piston with the finish line in sight. They pushed it home for twelfth place. On his return to Australia Jeffkins became a partner in the ill-fated Roo Motor Manufacturing Company.

Francis Birtles was an adventurer, writer and teller of some very tall tales. Nevertheless, he rode a bicycle around Australia twice and navigated for Syd Ferguson on the first ever crossing of the Australian continent from east to west by motor car (pictured). Birtles was on the road for nine months during 1927–28, becoming the first person to drive from England to Australia, yet when he arrived in Darwin from Singapore, his Bean 14 car was impounded by customs officials demanding import duties.

The father of the Holden car, Sir Laurence Hartnett, had been ingloriously demoted by GMH shortly before the first Holden car was unveiled. He left and launched a smaller, cheaper competitor called the Hartnett in 1949. Unfortunately it looked like something even Noddy would be embarrassed to drive. And neither this nor the later Lloyd-Hartnett (1957) provided a reason for GMH executives to lose sleep. That would come later, because Sir Laurence found an easier caper: importing strangely named new cars from a

Captain Arthur Waite won the race (later) recognised as the first Australian Grand Prix. That he beat Bugattis and other exotica in an Austin Seven suggests at least some luck was involved. The AGP became an international Formula One event from 1985.

Goggomobil production in Sydney. Be quick, you don't want to get an ugly one.

country few were taking seriously as a car builder: Japan.

The Goggomobil ('g-o, g-g-o' etc.) is a modern-day figure of fun but it was the only independently-produced local car of the 1950s to make money. Around 5000 Goggomobils were built in Australia, using a combination of local and imported parts. These included designs unique to Australia. The arrival of BMC's far more sophisticated Mini in 1961, however, effectively ended its days. The Zeta, built by the South Australian Lightburn company from 1963, stakes a claim as being the ugliest car ever built. And, as if to clearly emphasise how Lightburn should have stuck with washing machines and concrete mixers rather than cars, the company

AUSTRALIA JOINS IN

also bought the rights to locally assemble Alfa Romeos. The company's failure stretched from A to Z.

From the 1950s onwards it was the British car makers who set the pace when it came to going backwards. They started the decade with a massive share of the Australian car market (over 30 per cent for BMC alone) but managed to let that dwindle by two thirds by 1960. Such blatantly unsuitable 'pretend Australians' as the Morris Marshall contributed. It was an English transplant with a boomerang motif cynically stuck on the grille.

Only the wondrous Mini (let's overlook the Mini K, or Kangaroo, derivative) saved the Brits from largely disappearing when the Holden was joined by two other locally produced American 'compacts': the Ford Falcon (1960) and the Chrysler Valiant (1962). In the early 1960s BMC tried to launch its own local six-cylinder model, the Freeway, but it was another fiasco and local production ceased after only three years. By the early 1970s BMC had evolved into Leyland Australia and the disaster was even bigger: the P76 helped send the company down the gurgler.

Meanwhile, Holden went from strength to strength (some models were more popular than others, but all sold well). And with the exception of the first Falcon, which proved too fragile for local conditions, the locally-built Ford and Chrysler sixes and V8s were also well received throughout the 1960s. All became uniquely Australian by the early 1970s. But alas, not one home-grown attempt to build a fully Australian car ever came close to success.

> 'Love me, love my Holden. I laid that on the line with the missus before we were spliced.'
>
> Henry Williams in the Monaro-centric novel *My Love had a Black Speed Stripe* (1973). The book concerns, to quote the jacket flap, 'Ron's mad, passionate, consuming love affair with his car – an affair unappreciated by his wife, his neighbours and his boss.' Ron is forced to make a choice between his car and his marriage. Fortunately he makes the right one.

Many motorists refer to their car's paintwork as duco. In fact, Duco was a registered trade-name for a nitro-cellulose lacquer that was largely phased out during the 1960s.

FIVE COURAGEOUS ATTEMPTS

The go-nowhere car

The Australian-designed and built Caldwell-Vale powered and steered all four wheels before World War I. The vehicle also had four-wheel brakes and the steering system was almost certainly power assisted. Caldwell-Vale even termed its unique system 'all-wheel drive', a phrase that would not enter the popular lexicon until the 1990s.

Caldwell-Vale had built about 50 trucks before World War I, some with power-assisted four-wheel steering. The all-wheel drive passenger car was built on the suggestion of a buyer of one of these trucks and was first tested on the sand hills at Botany Bay, Sydney, in August 1913. Details of the engine and gearbox remain sketchy but newspaper reports at the time said the vehicle had a 30 HP engine and could climb steep dunes and take sharp corners at 60 km/h.

> 'They build the garage into the house these days – so the six-cylinder pig can sleep with the family, like in the middle ages.'
>
> Australian writer Barry Oakley in *Marsupials* (1979).

Caldwell-Vale's all-wheel drive tourer.

So the forward-thinking company went on to be a huge commercial success and a technological and industrial powerhouse – right? Not as such. Only one example of the extraordinary 'go anywhere' touring car was ever built and, metaphorically speaking, it quickly sunk up to its axles. The company was broke within a year of its completion, due mainly to a lawsuit concerning one of its trucks.

> More Australians died on the roads in the twentieth century than died in wars: about 160,000 in the first category versus the 89,950 killed in the four major conflicts Australia took part in.

Sincere Flattery

Although the Palm was advertised as 'ideally designed for Australian roads' and 'the car that all Australia is talking about', it hadn't been and the people weren't. Or, if they were even vaguely mentioning the Palm, it was as an alternative to buying it.

Beneath the flashy radiator and locally made mudguards, the 1917 Palm was a Ford Model T assembled from imported spare parts. The company behind it, Melbourne's E.W. Brown Motors, optimistically thought it could shift the Palm in big numbers despite it being almost twice the price of a standard Model T. And if slow sales weren't enough, Ford sued for patent infringement.

Palm production ceased but in the early 1920s E.W. Brown launched what it called an entirely new car. The 'Renown' had a nicely sculptured body and was boldly (and strangely) advertised as 'manufactured in the largest workshop in the world . . . by the Master Hand'. Again, most of what you couldn't see was Model T.

> 'My dad is 83 . . . He has always worked for himself, even in Lebanon. So when I got a job working for someone else – Ford – he thought I'd failed.'
>
> Lebanese-born Jac Nasser, speaking in 1999. Nasser, who grew up in Australia, had by that time become President and CEO of Ford Motor Company.

> 'So many Australians equate driving with masculinity; pass them and they suffer instant emasculation.'
>
> Author Ian Lawson Moffit in *The U-Jack Society* (1972).

The Renown sold as poorly as the Palm, and for all the same reasons. E.W. Brown's response was just a little predictable: another new car. This one was called the Spark, and you can probably guess the rest. Spark production ran to about ten units, compared with about 15 million for the car on which it was based.

The "RENOWN" Five-Seater

Showing the Smart and Sensible Hood

Note how beautifully the hood fits—how perfectly it follows the graceful curves of the superb "RENOWN" body. Side curtains are provided to safeguard the comfort of the rear seat passengers when riding in wind, rain or dust.

UPHOLSTERY DE LUXE

Hitherto the light car of moderate price has been sadly lacking in one great essential—comfortable and roomy seating.

In the "Renown" you are offered seating accommodation that is not rivalled in cars selling today at upwards of £600.

"Renown" upholstery is deeply filled, scientifically shaped and beautifully finished in Genuine Hide.

This taken in conjunction with the ample width of the body and the generous leg room provided, assures unsurpassably luxurious accommodation for passengers and driver alike.

The Renown, also known as the Palm, the Spark – and the Model T Ford.

Timber!

The car designed by dentist A. R. Marks around 1923 was different to most in that it was built entirely from wood. Okay, the mechanical components weren't timber but the use of stressed plywood throughout the body meant the vehicle had no need of a conventional steel chassis. This unitary construction technique was claimed to be exceptionally strong and therefore ideal for 'colonial conditions'.

A Sydney-made four-cylinder engine sat east–west across the chassis to optimise weight distribution. The rear wheels were driven by chain through an unusual transmission system that gave four forward speeds without a clutch pedal. Marks's son, Jim, discussed the unusual machine with Charles Kingsford Smith, a flying colleague. Kingsford Smith, then Australia's most famous aviator (and now an airport) was keen to become involved.

Wood-be contender: The Southern Cross and, below, Smithy (centre).

The 1922 Summit, built in Alexandria, Sydney, and advertised as 'an Australian Triumph', was possibly the first car in the world with a radio as standard equipment. Daimler also stakes a claim: a radio was fitted to its 45 hp limousine of October 1922.

> **'Could you move the Camira? I need to get the Torana out so that I can get to the Commodore.'**
>
> Typical request of tow-truck driving father Darryl in the typical Australian home of the Kerrigan family. From the film *The Castle* (1997).

In 1933 the updated, laminated timber car – renamed Southern Cross after one of the planes flown by 'Smithy' – made its debut amid much publicity. However, only a handful of Southern Cross cars had been sold by 1935 and Kingsford Smith went to England, apparently looking for more funding.

On the way back he thought he'd have a shot at the England–Australia air speed record. Unfortunately, he disappeared over the Bay of Bengal. The car-building project went no further. The one surviving example was reportedly burnt to a crisp in the 1970s by an angry wife, following a domestic.

Australian automotive colour chart atrocities from the 1970s included Bondi Bleach White (Chrysler); Candy Apple Red (Ford); Kanga Blue (Chrysler); Oh Fudge (Leyland); Monaco Maroon (Holden); Little Hood Riding Red (Chrysler); Strike Me Pink (Holden); Tan Fantastic (Chrysler); Home on th'Orange (Leyland); and Thar She Blue (Chrysler).

Too much, too early

It seemed a great idea: serve up a solid 'Australian Six' as a rival to that spindly American four, the Model T. Car buyers mindful of power and patriotism would be instantly gratified, as would the investors in this brilliantly conceived new venture. Er, wrong. On mature reflection, trying to take on the Model T Ford with what was essentially a kit car built in suburban Sydney was something less than a sure-fire, absolute cert, can't miss, mega-hit.

The story started around 1917 when Frederick Hugh Gordon visited the US and met Louis Chevrolet, then marketing a six-cylinder model against the all-dominating

Ford 'four'. Gordon was inspired and reasoned that, with a combination of imported mechanical parts and local assembly, he could do a similar thing down-under.

After teething dramas, Gordon produced a tally of 49 Australian Sixes in 1919 and met his £495 target price. At a time when imported sixes were selling at £750 and more, it was a bargain. The problem was that Gordon was losing about £2000 on each car he sold.

Production of the Australian Six – its distinctive radiator was guaranteed not to boil – climbed during the next few years. But, by 1925 the price of a Ford T had plummeted to around £185, while the Australian Six was now more than four times as much. Gordon was penniless and out of the business by then but production trickled on until about 900 examples had been sold. That total would stand as a record until the Holden came in 1948. But it had been a very expensive one to set.

A long-bodied Australian Six.

STEP ON IT!

> A 1963 EH Holden S4 will cover the standing 400 metres in around 19 seconds, making it quicker than a Boeing 747. From about 401 metres onwards the Jumbo is more impressive.

A shouldabeen

Victoria's Bolwell brothers – Campbell and Graeme – dreamed of producing an internationally successful Australian sports car. So did hundreds of others, and scores of them went as far as to build a car or two. The Bolwells did better, producing a string of models, each more accomplished than its predecessor and a little more numerous.

In 1967 the Holden-based Mark 7 coupe was launched to much acclaim. In late 1969 the Bolwells produced the even more spectacular Nagari. With its low, seamless fibreglass body, side gills and doors that cut into its roofline, the Nagari captured world attention. Built around a substantial backbone chassis, it was a full third lighter

Bolwell Nagari coupe.

Bolwell Nagari roadster.

than the Mark 7 and, as if that weren't enough to guarantee improved performance, it was powered by Ford's 4.9-litre V8.

The Nagari's height of just 93 centimetres meant that the cockpit was cramped, and the placement of the V8 right at the back of the engine bay (for better weight distribution) ensured a hot and noisy cockpit. But many big name exotics had worse problems than that and cost a great deal more too.

Things may have looked promising – particularly with the addition of the graceful Nagari Roadster – but the main problem was the usual one: economies of scale. Late in 1974 Bolwell announced that production of the Nagari was no longer viable. Perhaps it never really had been, but about 140 Nagaris were built and they are now much sought after.

STEP ON IT!

AND ONE SUCCESS

The first big-selling Australian car, GMH's 48-215 model, received its Holden badges only at the last moment. Names such as Koala, Stralia and Melba were considered, if only briefly, while Anzac, Canbra and GMH were short-listed.

Many executives argued for Chevrolet, believing the only way to persuade people to take an Australian car seriously was to pretend it was something else (not that it wasn't something else – although made only in Australia, the design was that of a discarded Chevy). What eventually, and unexpectedly, got the nod was the name and logo of the old Holden body-building division which the GM Corporation had

> 'The Holden is the best example of Australia's happy acceptance of second-hand Americana.'
>
> Australian architect Robin Boyd (1960).

The first Holden was tough and easy to drive, though not entirely in focus.

Car phones have been available in Australia since the late 1940s. Melbourne's 'PMG Manual Public Mobile Telephone System' of the late 1950s could service up to 50 subscribers. Unfortunately, only one of the 50 could use the system at any one time. The other 49 could listen in without restriction.

> The familiar 1948 Holden wasn't the first local car of that name. In 1911 one Roy H. Holden of Geelong, Victoria, built and put his name to a small, single-seater steam car. In 1915 there was an American Holden. The three wheels of this Texas-built vehicle had a layout that was somewhere between unusual and ludicrous: two were on the left of the car and one on the right.

bought in 1931 and merged with General Motors Australia.

On 29 November 1948, 1200 Australians joined their Prime Minister to ogle at the first official example: an ivory-coloured, six-cylinder sedan wearing the newly minted Holden badges. A ten-piece orchestra provided the accompaniment, while 44 radio stations were linked for the event. The fact that the best-selling car on the local market was then the Austin A40 possibly explains why everyone was so excited.

The Holden was quickly dubbed the 'Humpy' in honour of its podgy proportions. And although commentators used the term 'luxurious', there was one basic body, one level of

Later Holdens were much sharper.

> **Although production of all Leyland P76 sedans stopped in Australia late in 1974, the now-legendary automotive lemon continued to be assembled in New Zealand until early 1976, at which point the model was finally put out of its misery.**

specification and a choice of just four colours. The accessory list was similarly sparse: an 'Air Chief 5' car radio, a rear Venetian blind, a left-hand side sun-visor, a locking petrol cap and a heavy duty oil bath air-cleaner.

Still, the Holden featured Australia's first high-volume unitary construction body and with this light and strong shell it could meet its design target of '80 mph top speed and 30 miles per gallon' (130 km/h and 9.4 L/100 km). It was a sales winner from day one.

The FJ model that followed in late 1953 was a simple facelift. After the FJ secured Holden's position as Australia's favourite car, the FE, FC then FB appeared, though for reasons that mystify, the sequence wasn't followed with the FA (the next model was called the EK). Market share peaked in 1958 when the Holden outsold its nearest competitor two-to-one and GM cars captured just over 50 per cent of the total market. The millionth Holden was built in 1962, but the dream run was coming to a close.

If the 1950s and 1960s were times of sensational prosperity for Holden, the following decades were spent paying for it. As the market became more competitive Holden was left with too many factories, too many employees and not enough buyers. It survived the 1970s and 1980s by only the slimmest of margins.

CHAPTER FIVE

Maxim power

When the US industry journal *Advertising Age* published its list of the top 100 advertising campaigns of the twentieth century, the winner was a series of VW ads known as the Think Small campaign.

These much-celebrated ads, first seen in the USA in 1959, deliberately flew in the face of the bigger-is-better Detroit maxim. After starting out with a tiny photo of the Beetle above the banner 'Think Small', the series lampooned the Detroit habit of constantly changing its styling: a 1961 advertisement had a picture of the Beetle with the label: 'The '51, '52, '53, '54, '55, '56, '57, '58, '59, '60, '61 Volkswagen'.

The next year an identical ad appeared but the photo keyline was empty. The caption: 'No point showing the '62 Volkswagen, it still looks the same.'

VW's Think Small campaign.

STEP ON IT!

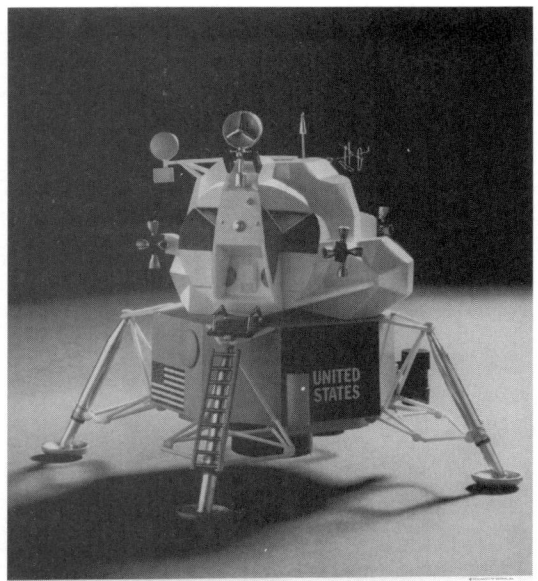

More of VW's Think Small campaign.

Volkswagen's 'non ads' succeeded in selling a small, low-powered, dumpy-looking economy car to high-income, well-educated Americans and showed even an ignoble profession such as advertising can have a golden age. Other automotive ads in the top 100 included the 1950s campaign 'See the USA in your Chevrolet' (number 41) and BMW's 'Ultimate Driving Machine' (number 84).

More than a century of mass-producing cars has involved more than a century of using slogans to 'shift metal'. They have ranged from the clever to the stupid, the meek to the

> When Australian F1 pilot Alan Jones bought his wife a Volvo he told the English press: 'She's not a very good driver and the Volvo is built like a tank. All it needs is a gun on the roof'. Volvo UK quickly produced an ad depicting one of its cars parked beside a Centurion with the line: 'The execution is different but the concept is basically the same'. To many eyes, however, the Centurion was prettier.

Left: Probably the world's first car ad: for the *Neu* and *Praktisch* three-wheeler from Benz & Co. The scantily clad women and jingles would come later.

Right: Racier approach: Mercedes-Benz puts forward its pitch in 1928.

shocking, and the informative to the blatantly untrue. In the earliest years of the century Packard introduced the catchcry 'Ask The Man Who Owns One'. It served the company for 35 years. 'Watch the Fords Go By' was one of the first and best known 'Henry' slogans, suggesting reliability and prolific production.

In the 1920s, Rolls-Royce was only one of many to announce its product as 'The Best Car in The World'. Yet on this occasion the maxim stuck. A later slogan, 'At 60 miles an hour the loudest noise in this new Rolls-Royce comes from the electric clock', became almost as famous. This was not the work of a copy-writer; it was lifted verbatim from a road test of the Silver Cloud that appeared in England's *The Motor* magazine. Considering the upright, wind-catching bodywork on the Silver Cloud, it must have been equipped with an exceedingly loud clock.

> **'It followed me home.'**
>
> A print ad (1990) quoting a buyer of the tiny Niki 650 explaining to the spouse the reason for his purchasing decision. Realistically, there could be no other excuse.

STEP ON IT!

> **'Two-thirds of the world is covered by water. Shame.'**
>
> The copy of a European BMW magazine ad (1997).

An advertisement – allegedly – for Jowett cars from 1936. The British brand is no longer in business. Can you guess why?

Barham and Gilbert

We have a very whole-hearted admiration of W. S. Gilbert, but we must say that, as a versifier, he couldn't hold a candle to the Rev. Barham. That is our opinion.

In one of Barham's efforts, "The Jackdaw of Rheims" we think, he delivers himself of the following, or thereabouts:—

"The sacristan he said no word to indicate a doubt,
But put his thumb unto his nose, and spread his fingers out."

A very rude gesture, but a gesture we feel like performing when we read certain advertisements.

Let us hope that no one will ever feel inclined to thumb his nose at us, "and to achieve this end" let us make a plain statement of fact.

Our car will not wash clothes (this should sell a fleet to Lever Brothers).

It will not sing the quintet from "Meistersingers."

It can't make 'em break from the off.

BUT, it can, and does, give more happiness at a cheaper rate than any other car made.

Verbum sapienti!

Will you have a Catalogue?

JOWETT CARS LTD., IDLE, BRADFORD

Local highlights include HDT's 'Body by Holden, Soul by Brock', Falcon's 'Trim, Taut and Terrific' and Valiant's 'Hey Charger' two fingers in the air salute. Mitsubishi's whisper-quiet Magna wagon was sold with line 'You'll Never Hear the End of It'.

In the USA in the 1950s and early 1960s, bigger was almost always considered better, which brings us back to those famous VW ads. Originally conceived in Germany, they had the most impact in the US, where they were further developed by the Doyle, Dane, Bernbach agency (President Kennedy and his Democratic Party were impressed to the point that JFK signed up VW's DDB to PR the DP).

Another example of VW's US approach: a blurred photo of what was unmistakably an entirely conventional garden-variety Beetle. The headline said: 'The X-93 experimental Volkswagen.' For a while Detroit scratched its head and added more chrome, then it started selling its own compact cars with deliberately austere styling.

After the success of the 'Think Small' advertising campaign, the Kombi was promoted with the slogan 'Think Tall'. VW's American TV ads of the 1960s and early 1970s were equally radical and self-deprecating. One showed a Karmann Ghia coupe trying to

Millions have been spent on single ads. These shots are from a famously expensive Chevrolet television commercial called Fusion in which an Impala magically assembles itself (and hopefully does it better than Chevrolet would). Forget computers: it was 1966 and each frame of film had to be hand-touched to remove the cables.

burst through a paper barrier – and failing. 'The most economical sports car you can buy', said the voice-over, 'just not the most powerful.' Not wanting to be left out, the Australian VW branch ran the copyline 'Volkswagen Loses' after the big Bathurst race of 1968.

A campaign is easier, of course, if the product is already a hit. In the mid-1960s the Ford Mustang was so successful that some US ads merely said: 'Mustang! Mustang! Mustang!' Conversely, Ford had to push the hard-to-shift Edsel with such lines as 'Dramatic Edsel styling is here to stay'. It wasn't.

The evangelical 'Datsun Saves' brought home the bacon for Nissan America during the oil shock of the early 1970s. The Bricklin Safety Sports Car (something of a misnomer

And there's always a stunt or two to get attention. The Rambler Javelin was promoted with the amazing Astro Spiral. By cleverly calculating speeds and ramp angles, drivers could complete a barrel roll in mid-air and land the Javelin back on its wheels. The trick was reused in the Bond film *The Man with the Golden Gun*.

considering the dud brakes, zero rear visibility and woeful handling) used the line: 'You'll think it's ahead of its time. We think it's about time'. An American Mercedes poster proclaimed: 'You wouldn't buy a cheap pacemaker. You wouldn't buy a cheap helmet. Let's talk about cars'.

MG used 'Safety Fast' in the 1930s, while 'Grace . . . Space . . . Pace' served Jaguar during the 1960s and beyond. Chevrolet boasted that its Corvette lacked something all other sports cars had: a role model. In the 1970s Holden ran with a ditty that supposedly defined all those things sacred to Australians: 'Football, meat-pies, kangaroos and Holden cars'.

The line credited with changing the downmarket image of motorcyclists in America was: 'You meet the nicest people on

VW was against using the name Beetle and only gave in when it was used almost universally by dealers and customers.

a Honda'. Volvo Australia used the opposite tack for its cars, boasting: 'You find them in the worst places'. The Swedish company's distinctive Australian ads have also used such slogans as 'Estate of the Art' (for the wagon) and before a state election: 'The safest seats in NSW'.

A very early Australian Volvo ad said: 'It's not everybody's car, but who wants to be everybody?' In 1995, a similar 'more individual than thou' theme was used in a British Audi television commercial. As a new A4 sedan is seen dashing through city streets, an excruciatingly irritating voiceover explains that winning is everything then mouths various other greed-is-good catch-phrases. At the end of the advertisement, the burke-in-question drives the Audi into a dealership, hands the keys back to the salesman and says: 'Thanks, but it's not my kind of car'. The screen cuts to the slogan: 'Audi: it's not for everyone'.

If one had to nominate the strangest motor campaign of

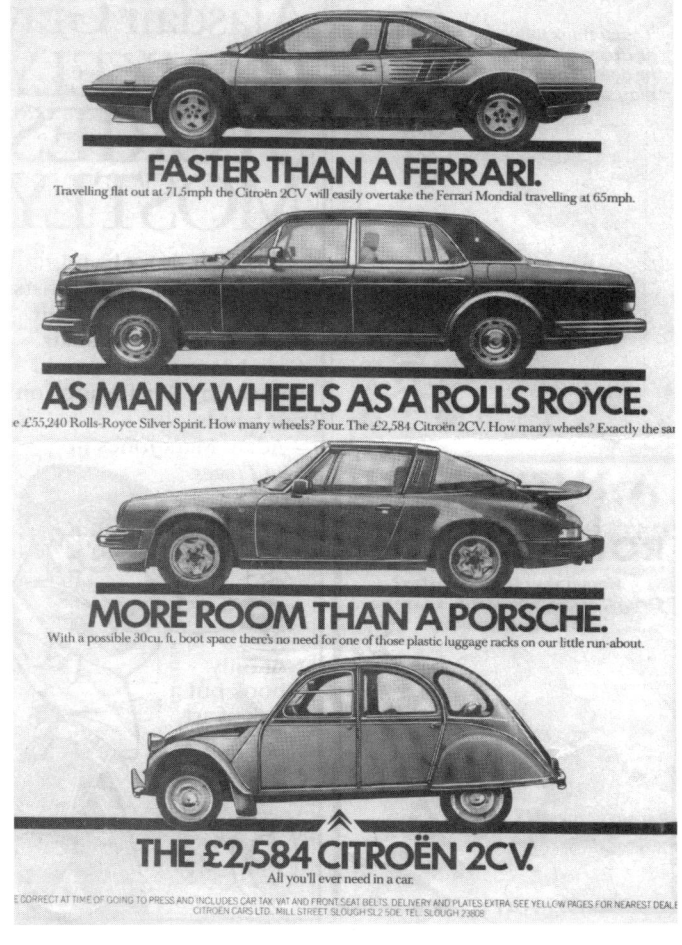

Citroën puts forth the case for the 2CV in the early 1980s.

STEP ON IT!

> Music has enormous clout in marketing, sometimes helping make a new model successful. Pop songs used in Australian ads have ranged from Cole Porter's 'I've Got You Under My Skin' (Peugeot 306) to Hot Chocolate's 'Everyone's A Winner' (Holden Vectra) and Bob Dylan's anthem of the obvious, 'The Times They Are A-Changin' ' (VW Passat).

'Save £28,000 today.'

The headline of a Triumph TR-7 advertisement that appeared in Britain in the 1970s. The claim was justified by showing the TR-7 coupe next to a Ferrari. The strange thing was that despite the price difference, it was Ferrari that had the waiting list.

all, it could well be the one Toyota used to sell its HiLux utility in 1999. The slogan – indeed the only two words in the whole HiLux television ad – were 'bugger' and 'me'.

Made in New Zealand, it starts with a farmer trying to right a leaning fence-post with the nose of his HiLux. He finds more power than he expected and – domino-like – flattens the entire fence for as far as the eye can see. His reaction is to utter 'bugger', a phrase repeated when a tree-stump he is dragging out with a towrope is suddenly wrenched from the earth. The stump sails over the ute and demolishes the outdoor dunny.

The farmer attempts to rescue a stranded cow with the

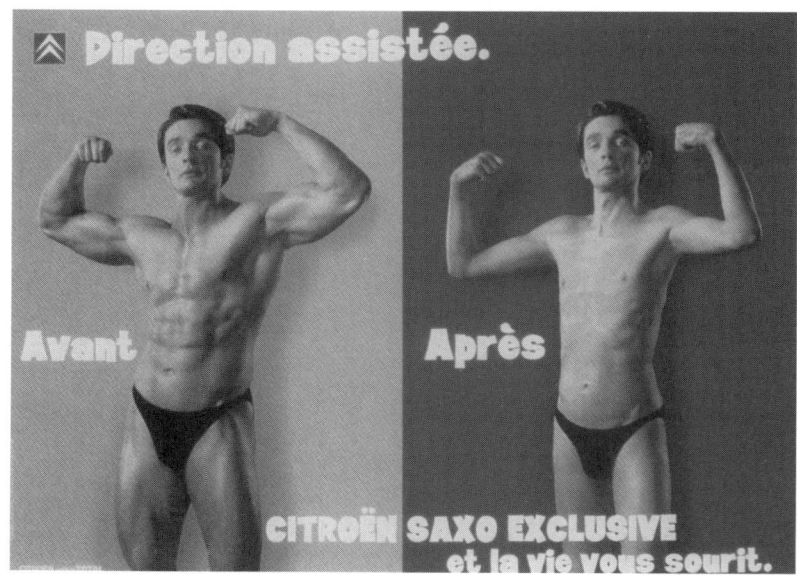

A typically eccentric Citroën billboard, snapped in Paris in the late 1990s. If you haven't picked it, this 'Before and After' duo are spruiking the optional power steering.

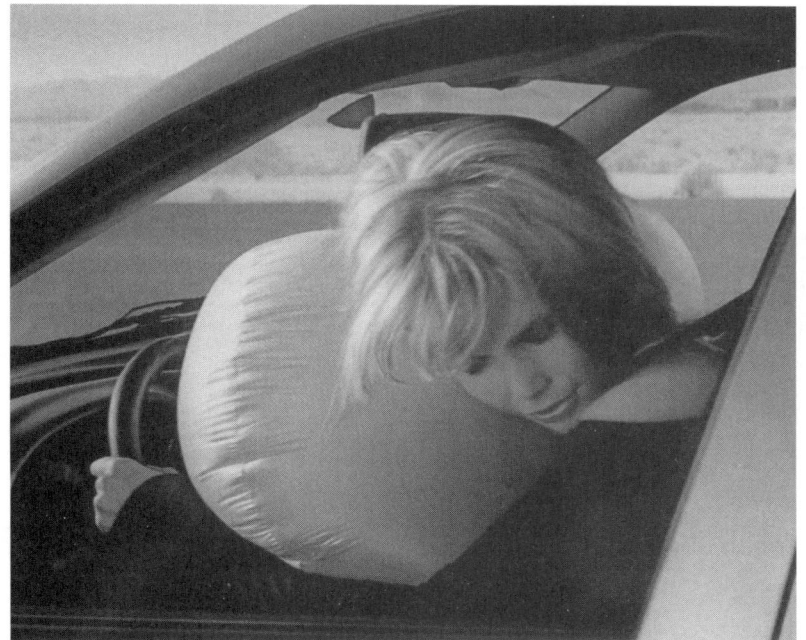

In a spectacular television ad, supermodel Claudia Schiffer sat in a Citroën Xsara during a real-life crash test, hitting the concrete barrier at 30km/h and brushing her heavily insured cheekbones against an airbag. 'I loved the concept of an advertisement with that element of danger', the real-life dummy was quoted as saying. What she exclaimed at the moment of impact was probably somewhat shorter.

same towrope. The results are not shown, though the strained bovine noises and the more sombre inflection on the next 'bugger' suggests all is not well. The final scene is of the farmer's HiLux spraying mud on the wife, the dog and the just-hung-out washing, leaving the wife to mutter the offending word. Thanks to computer animation, the dog follows suit.

Computer animation has also given us an Alfa driving through the canals of Venice, dancers morphing into a Eunos sedan, a Pontiac racing along the Great Wall of China, and a Citroën pulling itself to bits, reconstructing itself as a robot

> 'Almost anyone can achieve power, the trick is staying in control.'
>
> As the Eastern bloc was breaking up, Peugeot ran this line along with a picture of a 405 Mi16 sedan parked next to a toppled statue of Lenin (1989).

In the 1950s Studebaker produced a low-cost model with painted hubcaps, an austere interior and a penny-pinching level of equipment. Not a company to shy away from a national stereotype, it called the car the Scotsman. Production stopped in 1958.

STEP ON IT!

> In 1996 a certain German carmaker started using the song '(Oh Lord won't you buy me a) Mercedes-Benz' in its marketing, despite it being a satire of consumerism. It was the last song recorded by Janis Joplin before her heroin-and-alcohol fuelled death in 1970. All together now: 'Oh Lord . . .'

> 'In Italy, no one grows up wanting to be a train driver.'
>
> A European ad featuring a dramatic photo of the 20-valve turbocharged Fiat Coupe (1997).

and dancing. And today there is another new force: viral advertising. This involves small films being sent out anonymously by email, in the hope they will be forwarded by those who find them interesting, amusing or funny.

Such a technique took Honda's 'The Cog' advertisement around the world within hours. This awarding-winning ad, reputedly taking over 600 takes to get right, was set in a studio filled with car springs, pistons, doors and other components and created a domino effect that lasted a full two minutes. Each part moved the next along until the last snapped shut the boot-lid of a new car and caused it to roll down under a sign saying 'New Accord'.

Generally, the more controversial the viral ad, the quicker it spreads. None caused a bigger stink than a short suicide bomber film from 2005. It shows a man in a Yasir-Arafat-style checked scarf parking a VW Polo outside a crowded restaurant and pushing what looks to be a trigger for a bomb. There is a huge flash but the diners hear only a muffled blast; the car has completely contained the explosion, frying the bomber but leaving everyone outside to carry on as before. The tagline is 'VW Polo: Small but Tough'. Despite the high production values and the use of a new Polo, VW denied it was behind the ad and said it would find and sue those responsible.

The Polo viral that bombed.

84

Let's not forget the stupid press hand-out shot, another car industry mainstay. This natural-looking pose shows rallycross driver and skilled horsewoman Gill Fortesque-Thomas jumping 'her seven-year-old horse Royal Pheasant over her Wipac-sponsored racing Ford Escort Mexico'. Hold the front page.

Concentrated lows

- When Nissan launched – or tried to launch – its Infiniti luxury division in the States, it used a series of zen-inspired nature scenes. There was not a car in sight and the campaign's success was only with marketing textbook writers. They used it as the classic 'how-not-to'. As one Infiniti dealer expressed the problem: 'We're not shifting any cars, but sales of rocks and trees have gone through the roof.'

> 'You can bet that I will be in there pushing the '54 De Soto as hard as I can. I only hope I won't have to push it as often as I did my '52.'
>
> Groucho Marx, employed to promote De Soto vehicles in the 1950s, explains his intentions to the De Soto-Plymouth board of directors (1953).

> **'The car with everything also comes with the option of even more.'**
>
> A Holden brochure writer pays the price for exaggerating in the earlier pages of the Holden Vectra brochure (1997).

- With its 1989 slogan for the Maverick 4WD – 'It's different, it's a Ford' – an advertising agency managed to tell two whoppers in five words. The Maverick was merely a Nissan Patrol with a blue oval badge on the grille.

- An American Toyota advertisement from 1999 attempted to highlight Corolla reliability with the line, 'Unlike your last boyfriend, it goes to work in the morning.'. Problem was the ad appeared only in *Jet*, a magazine with a predominantly black readership. Complaints followed and Toyota apologised, saying that the ad was a mistake and acknowledging it could be considered offensive considering the high unemployment rate of black men in the US.

- In 2000 Mitsubishi Australia produced a too-bad-to-be-true TV ad in which a man's failure to perform in bed is rectified by stroking his Magna Sport and taking it for a fang.

- While Honda UK was winning advertising awards with 'The Cog', Honda Australia produced a television ad showing a 'depressed' car hurling itself off a cliff-top carpark because it wasn't as attractive as the new Accord. Branded 'grossly insensitive' by doctors and suicide prevention groups, it was withdrawn less than 48 hours after it was first shown, burning up about $400,000 in the process.

Why is Bridgestone, a Japanese company, blessed with such an English name? When Shojiro Ishibashi founded his country's first tyre company in 1931, he had his family name translated into English. The result was 'Stone Bridge'. He decided it sounded better the other way around.

> The Toyota company's first car was called a Toyoda, after the founder Kiichiro Toyoda. The first company president, Risaburo Toyoda, chose the altered name because it took eight brush strokes to write in Japanese rather than ten.

Kings of the road

It's not just cars that have left us with memorable motoring slogans. 'Put A Tiger In Your Tank', the catchcry of Esso in the late 1950s and early 1960s, produced hundreds of follow-up jokes and 'me-too' slogans. It also inspired a fashion for tiger tails hanging out of fuel flaps. 'Go Well, Go Shell' was famously used after World War II, while English chronicler Nigel Rees claims that, during the 1920s, a brand of petrol derived from tetraethyl lead used the slogan: 'Flat Out on Ethyl.'

Golden Fleece Petroleum once cheekily guaranteed its fuel would fit any shape tank. However, when it advertised 'The Big Gallon', it ran into trouble with authorities on the grounds it was 'amplifying a standard of measurement'. The famous 'Avis – We Try Harder' slogan developed from the controversial American slogan of the 1960s: 'When You're Only Number Two, You Try Harder. Or Else.' Admitting you were anything less than number one was not considered the done thing, and certainly not the American thing. But it worked.

Britain's Lucas Electrics used the slogan 'King of the Road' but the reputation the brand gained as standard equipment

> 'Well, burning a rolled-up Cuban leaf in your mouth isn't rational either.'
>
> The headline of an American magazine advertisement for Chrysler's modern hot-rod, the Prowler (1997).

> Sweden's Volvo company takes its name from the Latin term for 'I roll'. On its debut public run, the first Volvo rolled . . . backwards. The differential had been installed the wrong way.

in various English cars led to the suggestion that Lucas was an acronym of Left Us Cold And Stranded. The unofficial slogan became 'Lucas: Prince of Darkness', a line that, legend has it, first appeared in an obscure car club magazine; it was only a Lucas-initiated lawsuit which brought it to the attention of millions.

There were dozens of jokes – such as: 'Why do the British like warm beer? Because Lucas makes their fridges.' – and stories too that the company was about to build a vacuum cleaner: the first Lucas product that wouldn't suck. And then there was the wonderful line attributed to someone defending the company's products: 'I've had a Lucas pacemaker for years and I haven't had any prob–'

Smoking tyres

Monsieur Bibendum, the world famous Michelin Man, once smoked and drank. In his early days, the rotund one would sit at a table looking intense, immense and threatening, smoking a big Havana cigar and muttering 'Nunc est bibendum'.

The chain-smoking, glass-and-metal-eating Bibendum of the early days.

This Latin catchcry means 'now is the time to drink'. It's a line from first century B.C. Roman poet Quintus Horatius Flaccus, a man not quoted nearly enough in tyre commercials. What the human blimp was drinking was a champagne glass filled with nails, broken horseshoes and shards of glass. It was to emphasise that France's Michelin brand was the tyre that 'drank up obstacles'. In other early advertisements, the rubber man was seen as a giant, bashing up adversaries.

The 'Nunc est bibendum' slogan was dropped but the character itself became known as Bibendum. As for the cigar, early posters depicted Bibendum smoking it

After many years and at least one change of medication, Bibendum became the sort of big, fat, cheery rubber man you could take home to meet your mum.

through games of rugby, while motorcycling, while mountain climbing and even while courting. In 1929 the Michelin ad-artists were given the order to stop their tyres smoking. Nonetheless, Bibendum suffered a few tobacco relapses in the 1930s.

The original drawing was by the French caricaturist known as O'Galop in 1898. Over the years, Bibendum became more friendly in his appearance, the width of his tyres increased, his face became more defined and benign and nowadays he is never seen indulging in vices. Bibendum's success was a milestone in advertising, leading to hundreds of other companies creating mascots to give their products more personality. But few mascots have achieved the success of Bibendum, nor have they lasted the distance.

> **Legend has it that the famous Chevy 'bow-tie' logo was taken from a wallpaper design seen in a French hotel.**

CHAPTER SIX

Sharp curves

You've got to have style. Particularly if you're a car maker.

Style is what makes people pay twice as much for a svelte coupe as for an everyday hatchback, even when the two are remarkably similar under the skin. Style is what makes sedan buyers throw tens of thousands of dollars more at the seller to receive fewer doors. And style brings buyers into the new car market, people who were considering a second-hand jalopy right until the moment the television commercial showed the sweeping lines, the low slung grille, the plume of autumn leaves dancing behind the curvaceous new tail . . .

It wasn't always so. For the first 40-or-so years, it was considered enough that cars did what they did. Then came the 1927 La Salle, widely credited as being the first mass-produced car to be consciously 'styled'. The body and trim were the product of General Motors' new Art and Colour Section, arguably the world's first automotive styling studio.

La Salle was a cheaper version of Cadillac and also took

> 'Unique to the Limited were 15 utterly pointless chrome slashes down both rear wings.'
>
> A delightful comment from author Quentin Willson (*Classic American Cars*) when describing the 1958 Riviera, a bedazzling Buick with a grille boasting 160 separate protruding chromium squares.

'Car Number 9' was designed by Norman Bel Geddes in 1933. The former theatre stage designer was obsessed by aerodynamics so it was no surprise that his car had a streamlined body. Why it had eight wheels is less clear.

The Bugatti family was composed of artists as well as engineers. Car-builder Ettore's eldest son, Jean, designed the magnificent Bugatti Type 57 Atlantic. The famous crease down the centre was said to aid stability. Conveniently, it also allowed easy riveting of the two halves of the body.

The Facel Vega was an unusual French car in that most examples were powered by Chrysler V8s. Nobel Prize-winning author Albert Camus unwisely – and fatally – accepted a lift in one in 1960, even though he had already bought his train ticket. Camus had once said the most absurd way to die would be in a car crash.

STEP ON IT!

The 'coffin-nosed' Cord 810, first seen in 1936, was a sensation and not just because it was front-wheel drive. Sixty years later *American Heritage* magazine proclaimed it, 'The Single Most Beautiful American Car'.

> 'Tell me one thing Japanese technology has put on the automobile outside of coin-holders on the dashboard.'
>
> GM Corporation president Roger B. Smith (born 1925). He made the comment in 1990, his last as head of GM.

its name from a French explorer – in this case René-Robert Cavelier La Salle, who named Louisiana and was so popular with the men under his command, they murdered him. The man responsible for giving the 1927 La Salle its dramatic lines was a young designer named Harley Earl.

Earl became even more famous for his 1948 masterstroke: giving tailfins to the Cadillac Sixty Special. He had been inspired by the design of the Lockhead Lightning P-38 pursuit fighter. Cadillac's official history says: 'because there was so much excitement about this design innovation, dealers would often park cars in their showrooms with the backs facing the window, leaving the tail-lights on overnight.' The feature would soon spread to all other Caddies and across the market.

Earl was also responsible for the annual model update, arguing to his bosses at GM that 'noticeable change must come annually'. For this phenomenon he used the term 'dynamic obsolescence', which was very likely the origin of

> The Mini-Minor's designer, Alec Issigonis, defended uncomfortable seats on the grounds that they kept drivers alert.

1934 Chrysler Airflow 8. The public rejected it, even though it had done well in the first ever styling clinics. The company quickly produced a more conservative design and called it the Airstream, so as not to seem like it had stuffed up.

The first Saab was this 1947 streamlined two-door, two-cylinder, two-stroke prototype that looked like nothing else on the road. The proportions – somewhere between those of a slug and a radiation-proof bug – were not, alas, fully captured in the first production model, the 1949 '92'. That looked more like an armadillo.

The man who gave us fins: Harley Earl at the wheel of the 1951 Buick Le Sabre dream car.

> **'You can't make a silk purse out of a sow's ear, but you can make a very fast pig!'**
>
> Car modification advice attributed to legendary American car racer/builder Carroll Shelby (born 1923).

the less complimentary phrase 'planned obsolescence'. And, gee, didn't the idea catch on.

Other car makers realised there was more to be gained by changing what the buyer could see than what he or she couldn't. Today, car executives still say 'appearance and power are what sell a car – in that order.' VW was one of the few that stood alone in the 1950s, changing only the things you couldn't see.

Harley Earl held sway as GM's styling guru until he retired in 1958. He was replaced by Bill Mitchell, who styled the original Corvette Sting Ray, and was once quoted as saying 'chrome is my favourite colour'. So it was business as usual there.

Another master of the golden era of US design was suave, moustachioed French-born Raymond Loewy. He hated chrome, dubbing most US cars of the 1950s 'Jukeboxes on wheels, aesthetic aberrations that masked the workings of the machine beneath the layers of tawdry flash.'

Loewy designed cars, of course, but he also dictated how a great deal of America looked from the late 1930s through to the 1960s. The one-time *Vogue* fashion illustrator was responsible for toasters, refrigerators and household implements with simple outlines and rounded, streamlined looks. Loewy was responsible for Studebakers (including the Champion and Avanti), locomotives and Greyhound Scenicruiser buses. He even created the interior of Airforce One, the Lucky Strike logo and some of the most famous realisations of the Coca-Cola bottle (though not the original

---

In 1926 Oldsmobile staked a claim as the first car maker in the world to use chrome as a decorative device.

SHARP CURVES

Caddie lacks nothing: This Eldorado, photographed in 1957, featured air suspension, air-conditioning, power seat with memory, automatic door locks and a great deal more. The fins were gradually increasing in size too . . .

. . . reaching their zenith with the 1959 'King Fin' model.

Check out those Dagmars. From the 1954 model year, Cadillacs featured torpedo-like front bumper guards. These quickly became known as 'Dagmars', after the two unique selling propositions of Dagmar (Born Virginia Ruth Egnor), an otherwise forgettable blonde actress playing a nurse in *The Milton Berle Show*.

Stranger still, by the 1957 year model, the Cadillac Eldorado Brougham's 'Dagmars' had gained 'nipples'. Sigmund Freud would have had a field day – when he'd finished playing with the auto-opening and closing bootlid.

> **'A deckchair under an umbrella.'**
>
> Citroën chairman Pierre-Jules Boulanger sets the design parameters for the forthcoming 2CV model (1935). The onset of World War II delayed the launch until 1948. By 1950 there was a six-year wait. In 1990, when production stopped, about 7 million examples of the 2CV (and its derivatives) had been built.

contour bottle: Loewy is often incorrectly credited with this, perhaps because he had famously described it as 'the most perfectly designed package in the world').

Loewy is commonly known as the father of industrial design, partly because he kept as close a watch on the budget as he did on the form. 'The most beautiful curve,' he once said, 'is a rising sales graph.'

Another hugely influential designer of the era, and a man who had worked for both Harley Earl and Raymond Loewy, was Virgil Exner. Exner had a hand in various production Studebakers and the Chryslers, and produced some of the most outlandish 'dream cars' of the already-outlandish 1950s. Exner not only enlarged the fins pioneered by Harley Earl, he strongly argued they had a legitimate aerodynamic purpose. Which they did, as long as you were travelling at several hundred kilometres an hour.

The Europeans had a completely different approach to design, tending towards the functional end of the scale. However, they weren't entirely immune from American influence. Cars ranging form the Ford Anglia to the Opel Rekord and Mercedes 220S ended up with fins tacked onto their tails. Even the East German Trabant P601, aimed at the enemies of extravagance and consumerism, came to a

Raymond Loewy in his creation, the Avanti, in 1962. Unfortunately the company that was supposed to build it, Studebaker, was already in a death spiral.

conclusion with a pair of tailfins. In Japan the story was the same, as a look at the Toyopet Tiara or Nissan Cedric showed (mind you, a look at the Toyopet Tiara or Nissan Cedric was best avoided).

As fins increasingly came to be associated with American excess, the cleaner lines of European design studios started to have an impact around the world. By the later 1960s square and simple was the general rule everywhere except the US and France, which seemed to exist in a parallel design universe. France managed to produce cars that were as challengingly different (Matra 530) as they were ingeniously clever (Renault 12) and breathtakingly, gobsmackingly ugly (Citroën Ami).

It wasn't until the early 1980s that a near universal trend again emerged. This was the 'aero' design pioneered by Germany's Audi 100CD. With its rounded, 'organic' overall shape, flush-fitting glass, smooth mouldings and integrated grille, it looked ready to cut smoothly and silently through the air with a minimum of fuss.

> Germophobe Howard Hughes once owned a 1954 Chrysler New Yorker, complete with sealed windows and an air purification system that cost three times as much as the car.

> Mercedes pioneered fuel-injection as standard equipment on its 1954 300SL model. Chevrolet first offered fuel-injection in 1957 and turbocharging in 1962, yet spent most of the 1970s and 1980s turning out retro-tech tanks.

For the next 15 years most family cars designed and built around the world were heavily influenced by the aero school, so if you look at the original Audi now it looks remarkably conservative. In the second half of the 1990s body shapes started to sharpen with the influence of what stylists dubbed new-edge design (not just because it had new edges, but because stylists love to follow styles, and styles have to have names).

> 'A man who has once gotten automobiles into his blood can never give them up. A man with a dream can't stop trying to realise that dream . . . It's no disgrace to fail against tough odds if you don't admit you're beaten. And if you don't give up.'
>
> Preston Tucker, 1903–56, quoted in *American History Illustrated*. Tucker, whose controversial Tucker 48 (aka Tucker Torpedo) lost millions, never admitted he was beaten. Right up until his death he was trying to launch a new sports car project.

A 1951 example of the Studebaker Champion Starlight, credited to Loewy and known as the 'backward car'

GM designer guru Bill Mitchell with various GM concept cars. Unfortunately, the understated one couldn't attend.

Through all this, the place to find the most elegant and/or extravagant body designs in Europe was always Italy. The country had a long tradition of independent design houses and body-builders. Bertone, Pininfarina, Ghia, Frua and Giorgetto Giugiaro's Italdesign were among those that designed cars for fellow Italian companies and for car makers further afield. These companies shaped not only exotic supercars but economy sedans, big-selling family models,

Ford weighs in: the 1954 FX-Atmos.

> 'They're all image, packed with the symbolism of sex and power. They have the tails of rockets and chromium breasts like Jayne Mansfield's and when you hit the brakes, the rear end lights up like a robot animal in heat. Ultramatic ride, dynaflow penetration, triumph, lust, aggression and tons of room for the whole family.'
>
> Australian art critic Robert Hughes unloads on 1950s Yank Tanks in the television series *American Visions* (1996). But did Jayne Mansfield really have chromium breasts?

limousines and everything in between for companies from Japan, America, Australia, France, the UK and elsewhere.

The influence was such that Lorenzo Ramaciotti, general manager of Pininfarina, could say without irony in 2001 that 'Italian design has become a universal language of car design.'

And another: the 1955 Lincoln Futura was built by Ghia and attracted so much attention it was eventually modified to become the Batmobile for the high-camp 1960s *Batman* television series.

People complain that today's cars all look the same. In reality the best-selling models in each class have almost always looked the same. This is a Hillman Hunter and Ford Cortina Mark II. Or is it the other way around? Both models were built from 1966.

1950s excess gave way to . . . 1960s excess. The 1968 Chrysler 300.

Isn't beauty a subjective thing? This is a 1976 Lincoln Continental. Wonder where they got that grille from.

The Flying Lady, officially known as the Spirit of Ecstasy, was designed by sculptor Charles Sykes in 1910 and is still used atop Rolls-Royce radiators. The latest versions disappear in to the bodywork at the touch of a button to stop vandalism.

The French-made Voisin was capable of cutting, dicing *and* slicing pedestrians.

On the nose

In the early days, car radiator mascots were the only part of the vehicle that was 'designed' or 'styled', the rest being left to engineers. As cars became heavily sculptured, mascots were left behind or incorporated into less prominent badges. Only a few car makers still use mascots, Rolls-Royce and Mercedes-Benz being the most obvious examples. If interrogated on the subject, here's what you need to know:

- The origin of car mascots very likely rests with good luck charms on Roman chariots and terrets on working horses.
- The first manufacturer's radiator mascot was probably the one offered by the Vulcan Motor Company circa 1903. Fittingly, it depicted Vulcan, the god of metalworkers.
- Other early mascots mired in mythology included Unic's centaur holding a bow and Minerva's head of a goddess (Minerva was the goddess of the arts and professions). Gardner and Vauxhall both used a griffin, the legendary offspring of an eagle and lion.
- Ford used a bantam in flight for the Model A, though a miniature bust of Henry Ford was reputedly also available.
- Riley used a lady on skis, believing it epitomised 'smoothness and speed'. For much the same reason, Hispano-Suiza used the graceful Flying Stork of Alsace and Singer a gazelle.
- Pontiac used an Indian chief's head, Pierce-Arrow an archer, Packard a goddess.
- Alvis used a seated hare, a firefly and an eagle for different models.
- Hillman's winged bomb was remarkably prescient considering it was used from 1935 to 1938.

- Bugatti's famous Royale limousine of the late 1920s featured a rearing elephant on its enormous horseshoe-shaped radiator.

Jaguar's 'Leaper'.

If you can't have a mascot, at least have a distinctive grille. Get it right and you need never change it. But get the grille wrong and, well, it could look like this 1978 Chrysler New Yorker Brougham. Almost every grille trick has been tried, from none at all (such as the Studebaker Avanti) to networks of steel and chrome more elaborate than the bracing on the Eiffel Tower.

Girl power: the Galloway car (1921–1928) was produced in **Kircudbright, south-west** Scotland, by an almost all-female workforce.

Two BMW 'cars of the future' from the 1980s. Yeah, right.

Also slated for non-production: Honda's 1999 Fuya-Jo. It was designed to 'offer the playfulness of a skateboard' and had an interior supposedly inspired by a nightclub. And need we mention that it was unspeakably ugly?

Chrysler's Atlantic concept vehicle was an attempt to 're-imagine' the extravagant pre-war French coupe (even the name was pure Bugatti). Despite the warm reception when it was first shown in 1995, the Atlantic didn't make production.

Great moments in doors

Here are some cars that have dared to be different with doors:

Mini Moke: No doors, just a gaping hole and vinyl flaps that could be rolled up.

Goggomobil Dart: No doors and no gaping hole. The driver and passenger climbed over the side, leaving dignity on the footpath.

Messerschmitt KR200: Fighter-plane-style lift-up canopy, belying a complete lack of fighter-plane performance.

DeLorean: Gullwing doors proved graceful and functional on the Mercedes-Benz 300SL, but neither on this stainless-steel-bodied horror. The novelty wore off on the second use; the system usually broke down on the third.

Oldsmobile Toronado: American coupe with doors so huge and heavy its maker had to offer a 'built-in assist' mechanism to help struggling owners. Sagging was mandatory.

Lamborghini Countach: Doors opened upwards like a reverse scissor kick.

> **'I'm a Ford, not a Lincoln.'**
>
> US President Gerald Ford explaining how he differed from a previous White House occupant (1973). Pedants might note that, automotively speaking, a Lincoln *is* a Ford.

Lamborghini Countach: would suit quiet, retiring type.

Checker, famous for its New York cabs, produced this nine-seater wagon with six doors in 1962. There was a 12-seater variant too, with eight doors.

Alfa 156: Rear doors of this sedan had hidden handles so the owner could pretend it was a coupe.

Saturn SC1: This was a 'three-door coupe' from 1999 but the third opening wasn't a hatch. It was a short suicide door behind the driver's door. The handle was hidden in the jamb.

Mazda RX8: The Saturn idea, but on both sides: having two short suicide rear doors and no B-pillar created a 'four-door coupe'.

Rolls-Royce Phantom: The 2003 'BMW Rolls' introduced long suicide rear doors with conventional B-pillar to create a big four-door sedan with rear doors that opened the wrong way. Someone bought one, once.

The Ford Techna showcar from 1968. The doors opened parallel, which made it easier to enter and exit the car in shopping centre car parks, the designers argued. Wonder if they considered just making the car less than 3 metres wide.

SHARP CURVES

One problem with micro-cars is how to get in and out of them. The entire front of the BMW-Isetta bubble car opened like a fridge door, which was fitting as it was designed by Italian whitegoods company ISO.

Super-low sports cars make very little sense on the road. With the Bertone 'Stratos', the driver had to climb in first, so as to walk around the steering wheel. That black patch is a special 'walking platform'. Naff or what?

McLaren's F1 had gullwing doors that flopped forward rather than straight up. It didn't help the company find buyers.

The Winn electric car about to chomp down on its owner.

> Adolf Hitler admired Henry Ford and gave him the dubious honour of being the only American mentioned in *Mein Kampf*. Hitler also borrowed a novel Ford system of selling cars, whereby would-be motorists bought stickers, stuck them into 'savings books' and, when they had collected enough, exchanged them for a car. Ford used this to sell the Model T; Hitler later tried it with the VW. The difference was that those who put their faith in Nazis never received their cars.

SHARP CURVES

Silly doors at dawn: GM's stillborn Wankel-powered AeroVette from 1973 in front of the Lamborghini Marzal from 1967.

The Purvis Eureka's entire roof and glassware lifted up. Or didn't, if the mood took it. The couple in the background have given up trying to get in.

CHAPTER SEVEN

Cars on film

It didn't take long for the automobile to become a reliable film extra. After all, a car could instantly and cheaply add colour and movement. Or, in the early days, black-and-white and movement.

A car could also help define a character more quickly and precisely than a whole swathe of dialogue: an audience knew that what could be expected from a man who jumped out of a beaten-up Ford was completely different to what could be expected from one who leapt from a racy European sportster or a big black English limousine.

The first film break came when Charlie Chaplin, Buster Keaton, the Laurel and Hardy team and other masters of the one-reel silent comedy decided that automobile accidents were funny. From a car's point of view, it wasn't the most glamorous role around but, hey, it was still show business.

To many people, horseless carriages and movies were just fads, yet the car went on to play a big part in Hollywood

> **'It isn't simply stupid. It's trying-to-be-smart stupid.'**
>
> Pauline Kael, famously acerbic film critic from *The New Yorker*, reviews *Mad Max 2* (1982). The film was known as *The Road Warrior* in the US. The sparse dialogue included Max Rockatansky (Mel Gibson) saying 'I'm just here for the gasoline.'

dramas, as well as adding cuteness and light relief to Britain's Ealing comedies, providing action for Elvis's celluloid atrocities and techno-wizardry to the film exploits of James Bond and countless other heroes and villains. A prolific director of B-grade movies once proclaimed 'breasts are the cheapest special effects'. If not the second-cheapest special effect, fast cars are certainly among the most popular, particularly when one is pursuing another. The car chase was at first a plot punctuation but grew to the point where it was the plot that provided the punctuation to the car chase. *Bullitt* is often described as a classic movie, but who can remember anything about it other than the sliding, skidding, leaping vehicles?

The 1982 film *The Junkman* destroyed over 150 cars, while

The Car, from 1977, told the story of 'a mysterious, malevolent machine that terrorises a small US town without apparent motive'. It was a rip-off of *Duel* but added the suggestion that the devil himself was driving the machine. America's *TV Guide* said: 'Good score, OK crash sequences, and lots of unintentional laughs are the only reasons to sit through this movie.'

STEP ON IT!

Batman's various film appearances have involved a series of Bat vehicles, including this 1960s motorcycle, with Robin-carrying side-car. Now, now, if dressing like that makes them happy . . .

'Our Lady of Blessed Acceleration, don't fail us now!'

Elwood Blues speaks out in the John Landis-directed, Dan Aykroyd and John Belushi-starring film, *The Blues Brothers* (1980). In automotive terms, the film is memorable for the extraordinary gathering of vehicles at the end, along with innumerable wrecked cars during the chase scenes. A pair of Nazis drive a Ford Pinto station wagon.

the car chase – or truck chase – helped launch the career of one Steven Spielberg (*Duel*, 1971). *Mad Max* flicks made an art form of mutant machinery. *Chitty Chitty Bang Bang* and *The Great Race*, both from the 1960s, traded on the glamour of the early automobile age; *Chitty Chitty* was even reborn as an all-singing, all-dancing West End musical in the twenty-first century.

Motor sport, however, has generally fared badly on film. Tom Cruise's NASCAR pic *Days of Thunder* was lamentable, and the same goes for any number of motor racing biopics. And the word to correctly describe Sly Stallone's *Driven* (2001) is yet to be invented. The simplest explanation for the motor-sport-makes-for-bad-movies rule is that the real thing is available and pretty damn good, so Hollywood isn't required. Then again, you could say that about sex. Here are some films in which cars are stars.

Genevieve (1953)

The slight and gentle plot of this, one of the most loved of English comedies, centres on a pair of competitors staging a friendly race on the way back from the London-to-Brighton 'Old Crocks' Rally. Kay Kendall, Kenneth More and a load of others put in sterling performances. Although held up as the archetypal Ealing Studios film, *Genevieve* was actually made at Rank due to logistical problems at Ealing. Larry Adler played the famous harmonica theme, while a 1904/05 Darracq starred as 'Genevieve'.

The film *Genevieve* made veteran cars popular, though some continued to argue that if such cars were any good, they'd still be making them.

STEP ON IT!

A special Lotus Esprit submarine was created for the Bond film *The Spy Who Loved Me*. Arguably the first ever Lotus that didn't leak, it had been modified by Florida firm Oceanographics Co. and was powered by four electric motors. It could supposedly travel at 7.2 knots at a depth of 160 metres and – in the film at least – fire missiles to ward off hostile underwater traffic.

On The Beach (1959)

Set five years in the future (1964) and based on the book by Nevil Shute, *On the Beach* details how, with the world largely destroyed by nuclear war, civilisation exists only in Melbourne. That's not the only temporary suspension of disbelief required: an ancient Fred Astaire wins the '1964 Australian Grand Prix' – in a sports car. The race is supposedly at Phillip Island, and no-one is too concerned about surviving because the radiation is going to get you if the brake fade doesn't. Gregory Peck and Ava Gardner star, while all the Australians go fishing and sing *Waltzing Matilda* for what seems like hours. Daft.

Comedian Peter Sellers once ran an advertisement in *The Times* stating 'Titled motor car wishes to dispose of owner'. He was divesting himself of a 1958 Silver Cloud, soon to be replaced by a Bentley Continental.

The Green Helmet (1961)
This black-and-white English film is one of quite a few low-budget films of the 1950s and 1960s featuring real race-drivers, as opposed to actors playing car drivers. Jack Brabham, Roy Salvadori and a few lesser lights play themselves alongside, wait for it, Sid James.

Spinout (1966)
'It's Elvis with his foot on the gas and no brakes on the fun!' said the promo line. To put it more realistically, it was another formula film from the Pelvis with little to commend it. Elvis plays a singer and part-time race-driver. Even worse was *Speedway* (1968) where he plays a driver who is successful but has tax problems. Anyone still awake at the end should have been given a trophy and allowed to pop a magnum of bubbly.

James Bond's Aston-Martin Volante, from *The Living Daylights*. Although Bond is most commonly associated with Aston-Martins, he has driven plenty of other vehicles in his time, including a 1962 Sunbeam Alpine in *Doctor No* (Sean Connery), a Willard Whyte Techtronics Moon buggy in *Diamonds Are Forever* (also Connery), a Renault 9 taxi in *View To a Kill* (Roger Moore) and a Soviet T72 tank in *GoldenEye* (Pierce Brosnan).

> **'Because I don't believe everything I read.'**
>
> Boston comedian Steven Wright explains to a Sydney audience why he drives through stop signs (2005).

Grand Prix (1966)

This visual delight has some of the best 1960s motor racing footage ever captured, much of it shot during real Grands Prix. Ex-world champ Phil Hill drove the camera car and Graham Hill, Bruce McLaren and others made guest appearances. Eva Marie Saint and James Garner are the big name actors. Françoise Hardy looks magnificent (in contrast to the way she acts), while Yves Montand plays the great Italian driver 'Sarti' in a Maserati which is playing a Ferrari. There's the usual lack of plot but the dialogue is sometimes sharp and director John Frankenheimer uses all manner of special effects in the racing scenes.

The Love Bug (1968)

Disney had a huge hit with this tale of a VW Beetle that has a mind of its own and garners great success on the racetrack. It spawned such sequels as *Herbie Rides Again* (1974), *Herbie Goes to Monte Carlo* (1977) and *Herbie Goes Bananas* (1980)

The Love Bug and its star, Herbie the VW, were first seen in 1967 and rode again, and again and again.

About 500 VWs showed up at T*he Love Bug*'s Sydney premiere in at Chullora drive-in.

in roughly descending order of quality and interest. The franchise was reborn in 2005 with *Herbie: Fully Loaded*. The NASCAR-centric twenty-first-century remake featured a New Beetle in the plot, but the main car was still a 1960s Bug with the number 53 painted on its doors.

The Italian Job (1969)

From the opening scene featuring the Lamborghini Miura (soon to be a steaming wreck), to the fabulous jumping Minis, this just might be a car-lover's perfect film. Every Brit who is any Brit, including the just-about-to-be-knighted Noël Coward, gets involved in this chin-up, stiff-upper-lip, let's give the Wops what-for extravaganza. French stunt drivers – none of whom had driven a Mini before the film –

STEP ON IT!

did the steering, and proved Minis can leap 25 metres between buildings, slosh through drainage pipes and go down a flight of stairs quicker than Princess Di. The final scene (with a precariously perched bus) was set-up for a sequel that was never produced. Pity.

Winning (1969)

This was the film that inspired its male lead, Paul Newman, to go racing for real. He took up the sport seriously in 1972, drove to second place at Le Mans in 1979 (aged 54) and

> Film director Roman Polanski bought his actress wife, Sharon Tate, a white Rolls-Royce Silver Dawn shortly before she was murdered by members of the Manson cult. Roman gave his own red Ferrari to Sharon's father after the murder.

CARS ON FILM

This spread:
Was there ever a better car movie than the original *The Italian Job*? The Mini escape vehicles weren't the only car connection. The crooks steal $4 million worth of gold the Chinese have flown in to Turin as a deposit on a Fiat car factory. The stunts destroyed ten Mini-Minors, five Alfas, four Jaguar E-Types, two Aston-Martin DB4s and – shed a tear – a Lamborghini Miura.

became the co-owner of the Newman–Haas Indycar team. The film itself, which incorporates footage of the 1967 Indy 500 interspersed with the usual car-movie love affair (Joanne Woodward co-stars), is not nearly so interesting. Famed British critic Leslie Halliwell judged *Winning* 'well but needlessly made'.

The Burglars (1971)

Not so much a film as a car chase with padding, this dubbed French/Italian effort starred Omar Sharif and Jean-Paul Belmondo (whose real-life son was later to briefly race in Formula One). During a hectic and extraordinarily long pursuit, Belmondo drives a mysteriously self-repairing Fiat 125 sedan; Sharif is in an Opel Rekord. The plot is obviously in a Morris Marina.

Le Mans (1971)

Filmed on location during a certain 24 hours, Steve McQueen's *Le Mans* contains some engaging motor racing footage, but a rounded film it is not. Indeed *Le Mans* is such a plot-free zone it's hard to believe it was based on a novel. Burgess Meredith narrates, while Jacky Ickx, Derek Bell and others (including McQueen) drive especially for the cameras. *Le Mans* cost a bomb to make but earned three-fifths of not very much. McQueen's bank account was severely wounded, the actor having unwisely pitched in when the film went hopelessly over budget.

In the Humphrey Bogart–Ingrid Bergman film *Casablanca*, two of the minor characters were named Ferrari and Renault. Viewers and critics later presumed this to be in-joke about two brands of car. However, the film was from 1942 and the Ferrari car-making concern was not formed until after World War II.

CARS ON FILM

> The reason Roger Moore drove a Swedish car, a Volvo P1800 coupe, in the famous 1960s English television show *The Saint* (based on books by author Leslie Charteris) was that Jaguar couldn't or wouldn't supply an E-Type by the time filming was ready to begin. Volvo contacted the TV production office, which agreed to pay full retail for a white P1800.

The Cars That Ate Paris (1974)
Peter Weir directed this neo-Gothic something-or-other, while a bunch of weird-looking cars starred alongside Terry Camillieri and John Meillon. Paris was an Eiffel-Tower-free town in western New South Wales, where the residents made a nefarious living pilfering the cars of visitors. At the end, a fleet of motorised monstrosities wrecks the joint. Is it deep? Hard to say.

The FJ Holden (1977)
Drunks, drag races, love gone wrong and, of course, a beloved FJ. This film looks and sounds like the result of someone walking around Bankstown Square shopping centre in Sydney's west with a Handicam, if indeed they had such things in 1977. A young Sigrid Thornton plays Wendy, while Paul Couzens is Kevin, apprentice mechanic and owner of the FJ. Jim Manzie of the band Ol '55 wrote the score but that group's original singer (the appropriately named Frankie J. Holden) didn't appear.

Mad Max (1979)
This tale of a future wasteland dominated by gangs and malevolent-looking modified cars made a star of Mel Gibson; ironically, the American-born actor mumbled so much his voice was re-dubbed for the US version. *Mad Max 2* (1981)

> 'But you've got to be sober to fly . . . I mean it's not like driving a car.'
>
> Homer Simpson explains the facts of life to his permanently soused friend Barney Gumble, who has decided to take helicopter lessons. From *The Simpsons* ('Days of Wine and D'ohses' episode, 2000).

Mad Max madness: a 1959 Chevrolet Impala is hacked up by the Toecutters Gang. . . . and Max's Ford Falcon XB pursuit car tries a spot of caravanning.

was a bigger-budget remake rather than a sequel. The cars were still mainly recognisable Holdens and Falcons, while *Mad Max Beyond Thunderdome* (1985) put a lot more work into making the vehicles look as misshapen as the characters.

> The late 1960s cartoon television series *The Wacky Races* introduced the character Peter Perfect, a name later associated with nine-time Bathurst winner Peter Brock.

Roger and Me (1989)

Michael Moore (later of *Bowling for Columbine* and *Fahrenheit 911* fame) wrote, directed and appeared in this highly entertaining but rather partisan (and by some accounts, fib-filled) documentary. The plot concerns the disgruntled film-maker's attempts to meet Roger Smith, then GM CEO, and explain to him that the town of Flint, Michigan, shouldn't have to suffer because of another GM restructure.

The Big Steal (1990)

Not to be confused with the 1949 US film starring Robert Mitchum, this Aussie effort presented the highly unlikely scenario of a used car dealer doing things that are dishonest (the dealer is named Gordon Farkas and is beautifully played by Steve Bisley). Ben Mendelsohn comes of age for the 100th time, Claudia Karvan co-stars, as does a Jaguar XJ6 and a Nissan Cedric.

The Italian Job (2003)

Hollywood's arrogance has rarely been shown to greater effect than with this ugly and unnecessary remake that takes an archetypical British film, albeit one set in Italy, and moves most of the action to Los Angeles. The new BMW-built Mini replaced the original. Plenty of action, zero charm.

> **'DONOR CYCLES:** term for motorcycles, because accident victims make good organ donors.'
>
> From a list of authentic US medical slang published in the book *Behind the Scenes of ER* (Warner Bros, 1996).

The reborn Mini looked the part in the reborn *The Italian Job*, but the film let it down.

Fast meeting

C'était un Rendezvous lasts less than ten minutes and has no plot, no dialogue and no music. Yet there is a reason it has become a cult classic: it consists solely of an attempt to drive a Ferrari across Paris at the fastest possible speed, with a low-mounted camera catching all the sounds and sights.

Insane speeds mix with ridiculous risks as the mystery driver hits 180 km/h down the Champs Elysées at dawn, rockets through Place de la Concorde and past Opéra Garnier before screeching through narrow, cobbled streets up towards Sacré Coeur at Montmartre. Along the way, the driver blasts through endless red lights, slices between garbage trucks, buses and 2CVs and takes the viewer on a real-life ride more dramatic than any video game.

The film is steeped in mythology and for many years the only available copies were bootlegged VHS tapes. Stories have been rife and varied about who made the film and which motor racing superstar was behind the wheel. In reality, the unlikely man behind *Rendezvous* was Claude Lelouch, the director of the 1966 French romantic hit *A Man and a Woman*.

It was left to a British enthusiast, Richard Symons, to locate an original 35mm negative print and have the film remastered and transferred to DVD. In the same year, 2003, Symons coaxed details from Lelouch, who confirmed it was himself and not any of the rumoured French Formula One pilots who drove at

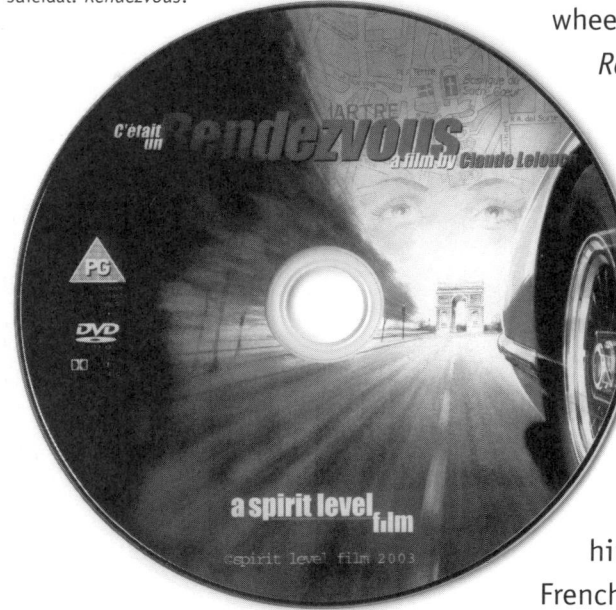

Short, sharp and near-suicidal: *Rendezvous*.

such frightening speed (he wouldn't reveal, however, what type of Ferrari it was or, if indeed, it definitely was a Ferrari).

'The film lasts nine minutes and 30 seconds – precisely the same amount of time it took to shoot,' Lelouch explained at the time the DVD was released. 'This is because it's exactly how much I had left after filming *Si C'était a Refaire* [a.k.a. *If I Had To Do It All Over Again*, starring Catherine Deneuve]. At the time I thought it would be a shame to waste those 300 metres of celluloid. So I took the opportunity to create another film that had been close to my heart for some time.

'So I have 570 seconds, not one more, to do Porte Dauphine to Place du Tertre [with] the impossibility of guaranteeing the safety of the operation. I have limited the risk by filming in August at 5 am. But I have not been able to close down the streets on my route.

'The most dangerous part of the drive is the passage at the gates of the Louvre. There is no view of the exit – if a car appears at that moment a collision is inevitable. I strategically placed my assistant, Elie Chouraqi, there with a walkie-talkie, so he could advise me of any danger. When I arrived level with the gates there is no signal, so I step on it. The rest of the journey is accomplished ... 15 minutes later I find Chouraqi fiddling with his walkie-talkie. I ask him, "What is it?" and he says, "It's this piece of shit!" pointing at it. "It stopped working at the beginning of the take!" A shiver goes down my spine.'

Although some have expressed doubts about the soundtrack of *Rendezvous* (the engine sometimes seems to be accelerating when the car isn't) and some believe the film has been sped up (Lelouch denies it), the overall appeal is easy to see. It was filmed 'live', it is daring, it is dangerous, it is as un-PC as you can be.

> 'I do have four Cadillacs, but I haven't got use for four. Maybe some day I'll go broke and I can sell a couple of 'em.'
>
> Elvis Presley talks to a radio announcer, circa 1955. Later Elvis had one of his many Caddies restyled to order by Barris Kustom City of North Hollywood. All the accessories – including bumpers, hubcaps and interior fittings – were coated in fourteen-carat gold and the equipment list ran to shoe buffer, record player, television, dual phones and gold lamé drapes. In 1966 Elvis's manager, Colonel Parker, arranged for this car to go on tour instead of the singer, who was too busy making another bad formula film.

STEP ON IT!

Perhaps to ensure there was at least one Smart thing in *The Da Vinci Code* film, director Ron Howard organised a car chase through Paris. The quirky little Smart Fortwo had also featured in Dan Brown's kazillion-selling novel on which the film was based.

Eight people from the arts and entertainment world killed on the road:

• T. E. Lawrence (of Arabia) (war hero, author, while riding Brough Superior motorcycle, 1935)

• Margaret Mitchell (Author of *Gone With the Wind*, pedestrian, 1949)

• Eddie Cochran (Rock-star and pioneer overdubber, while passenger in London hire car, 1960)

• Françoise Deneuve (Elder of acting Deneuve sisters, while driving Renault R8, 1967)

• Jayne Mansfield (Pneumatic actress, while passenger in Buick Electra, 1967)

• Harry Chapin (Folk singer and Ralph Nader supporter, while driving Volkswagen Rabbit, 1981)

• Tupac Shakur (Gangsta rapper, shot while sitting in BMW 750iL, 1996)

• Linda Lovelace (born Linda Boreman; Porn star turned anti-porn activist, while driving unspecified brand of car, 2002)

Entering the language

'What's in a name?' a bard once asked. 'Would not a car by any other name still get us from A to B. Or not to B, that is the question.'

But enough of that. If you are looking for alternative names for car, auto or automobile, there are plenty to choose from. *The Thesaurus of American Slang* by Dr Robert L. Chapman (Collins Publishers, 1990) lists 44, namely: Ark, banger, beater, boat, boiler, bucket, bucket of bolts, buggy, bus, buzz-buggy, buzz-wagon, cage, chariot, clunk, clunker, cochecito, crate, four-wheeler, gas-guzzler, goat, grinder, heap, hoopy, iron, jitney, job, junker, junk-heap, lemon, lizzie, lunker, puddle jumper, ride, set of wheels, sheen short, strugglebuggy, trans, transportation, tub, tuna wagon, vet wheels, winter rat, wreck and zoom buggy.

However, Dr Chapman's list is really only scraping the top layer of the duco because there are many, many other slang terms in popular use. Here, complied from various sources (including Australia's *Macquarie Dictionary* and *Macquarie Thesaurus*), are 100 or so more:

Battlebuggy, beast, beetle (VW), big Henry (a Ford), black maria (police car), black taxi (Commonwealth limousine), bomb, brick (original Mini), broom broom, buckboard, bug (VW or Bugatti), bus bait, bushbasher, buzzbox, chaffcutter, clanger, clapper, coffin, crock, cruise mobile, cruiser, death-trap, dog, dog cart, donk (usually just for the engine, but in some regions for the entire car), dunny-door (Holden Commodore), fangmobile, fizzer, flivver, freeway flier, gig, gin palace (station wagon), go fast, gritter, grunter, guttercrawler, hack, haroldwagon, heap, hearse, hot hatch, humpy (early series Holden), import job, jalopy, Jigger, kiddy cab, limo, LRB ('little rust bucket'), Marrickville Mercedes

> 'Throughout *Crash!* I have used the car not only as a sexual image, but as a total metaphor for man's life in today's society . . . I would still like to think that *Crash!* is the first pornographic novel based on technology.'
>
> J. G. Ballard describes his bizarre 1973 book *Crash!*, in which a group of car accident aficionados find ever more elaborate ways to smash themselves up. It was later turned into a suitably ugly film by David Cronenberg.

> 'Control is an illusion, you infantile egomaniac. Nobody knows what's going to happen next: not on a freeway, not in an airplane, not inside our own bodies and certainly not on a racetrack with 40 other infantile egomaniacs.'
>
> Dr. Claire Lewicki (Nicole Kidman) gives Cole Trickle (Tom Cruise) a piece of her mind in the NASCAR film *Days of Thunder* (1990). Tom's less-than-immortal dialogue in this less-than-immortal movie included: 'Speed! To be able to control it... to know that I can control something that's out of control!'.

(Valiant), matchbox, mechanic's nightmare, missile, 'mobile, mobile misery, mojo, motor, oil-burner, ol' box, old bomb, ol' girl, paddockbasher, panel van (aka carnal car, shaggin' wagon, sinbin and f**k truck), petrol wowser, pig, plagon wagon, raceabout, rattletrap, rice-burner, rocket, rod, runabout, rust-bucket, screamer, shaker, shandrydan, shitbox (specifically old or underpowered cars), shopping trolley, side banger, smoker, smokie, soap dish, square wheels, stayer, steamer, steel horse, street machine, tank, thumper, tin Lizzie, tojo (Japanese 4WD), Toorak tractor, truck, trap, ute, wheelbarrow, wheels, woe on wheels, workhorse, Yank tank, you-beaut (utility).

The industry also has its own language covering parts and sales techniques. When America's *Automotive News* used the term 'de-contented' in reference to the Chrysler Pacifica in 2005, the newspaper didn't mean the model was unhappy. It referred to a lower specification, or de-contented, version being considered by the company.

One industry masterstroke is 'Mandatory Option', a slick profit booster that works like this: a car can be specified with or without, for example, antilock brakes. Except, of course, that the manufacturer or importer won't sell it to you without antilock brakes, because they are a 'mandatory option'. Not surprisingly, they cost extra. There's the equally strange Delete Option. This means that leather trim, for example, is standard equipment at absolutely no extra cost. That is, unless you don't want it, in which case you can specify cloth and pay a lower price.

CHAPTER EIGHT

The boom time

After World War II almost every country except the USA was beset with stringent austerity. Rationing was still widespread in Europe well into the 1950s, forcing people to queue for meagre allotments of bread, milk, eggs and, crueller still, petrol.

Having bashed themselves and each other into submission during seven years of war, France, West Germany, Italy and No-Longer-Quite-So-Great Britain desperately needed export dollars to rebuild. They set about supplying manufactured goods to anyone, anywhere, who could stump up with the cash. And since you couldn't export houses, the next best thing was the second-most expensive item most people were ever likely to buy: a motor car.

To succeed in such lean times, these new European machines needed to be smaller than US models, with more frugal engines and body packaging that kept the exterior size down but the interior room still acceptable.

> 'What is good for the country is good for General Motors, and vice versa.'
>
> Everyone more or less knows the quote – but who said it? It was Charles E. Wilson, in 1953. Wilson had been head of General Motors and a Congressional Committee was asking whether his new position as Secretary of Defense represented a conflict of interest.

The French redefined the word sparse with the original 2CV. They gave the word ugly a bit of a rev-up too.

The cars that emerged – including the Citroën 2CV, the Morris Minor, the from-the-ashes VW and the reborn Fiat Topolino – put the emphasis on low purchase price and fuel efficiency. There was daring individuality on display too, and often profound ugliness as each maker found a different way of achieving the new goals. Styling was a luxury usually a long way down the priority list.

England was an early leader, quickly picking up its pre-war position as the world's number two car maker behind the United States. Forced to 'export or die', cash-strapped British makers opted for the former and, fortunately, buyers around the world were won over by the small, well-designed and cheap-to-run designs coming from Austin, Morris, Hillman and many other companies.

As a handy sideline, British sports models also became hot

In 1950 the USA accounted for 76 per cent of the world's car production. That figure had fallen to just 30 per cent by the start of the 1980s.

The first production version of the 2CV was much better equipped than the prototype. For instance, it had a second headlight. The doors, which they had forgotten to screw on to this example, were probably the thinnest and flimsiest ever made. In the same spirit, the boot lid was a canvas flap hanging down from the equally canvas roof. The seats were canvas too.

items. Light and low-cost roadsters from MG, Triumph and others were huge hits even in the US, and even when the MG was available to the Yanks only with the steering wheel on the wrong side. Jaguar's two-seaters and saloons were highly desired further up the price scale almost anywhere cars were sold, while at the very top, Rolls-Royce was still Rolls-Royce.

Germany quickly graduated from making some of the slowest and most spartan cars on the market (including econo-buckets such as DKW two-strokes and Messerschmitt three-wheelers) to some of the fastest and most luxurious. Italy and France too were able to gradually move upmarket

In the US extravagance reigned. 'Rear fenders sweep upwards and backward,' boasted the guff on this 1957 Dodge Lancer Hardtop, 'to keynote the car's swept-wing styling, accenting a sport car appearance that is also a mark of the jet aircraft age.'

> 'I saw an advertisement in *The Times* saying "sports car company for sale" . . . and I bought it.'
>
> British tractor maker and industrialist David Brown uses understated terms to explain his purchase of the Aston-Martin car company in 1947. Brown soon bought another famous British brand, Lagonda. His initials are still part of Aston-Martin model designations, despite the company being owned by Ford.

while still covering the people's car market. The Ferrari brand was born in 1948 and Citroën's revolutionary DS sedan arrived in 1955.

While this was happening, America's 'Big Three' car makers became increasingly insular, constantly restyling rather than reengineering, increasing engine size rather than efficiency. The US market was growing so quickly its car makers scarcely felt the need to cater for Americans on a tight budget, let alone foreigners who were skint. The result was a new generation of massive US 'land-yachts', with gadgets galore and gargantuan engines.

Curiously, many of these immense vehicles had no more passenger legroom than their smaller European counterparts. Furthermore, because of the more challenging driving conditions on the Continent, most of the Euros ended up with more sophisticated suspension designs and smarter transmissions than American cars too. Little surprise then that most of the worthwhile technical advances popularised during the post-war period (with the exception of the

Or you could have a Buick Century with a '255 horsepower V8 engine, with four-barrel carburettor, hooked up to the new variable pitch Dynaflow transmission'. It also came with 'fully exposed rear wheel', 'slanting doorbelt line' and 'distinguished new front end styling'.

automatic transmission) were from European makers. Three-point seat-belts (Volvo), disc brakes (Jaguar), antilock brakes (Jensen and Mercedes-Benz) and the Mini's front-drive/east-west engine layout are but a few examples.

Volvo claimed its three-point seat-belts, announced in 1959, led to a 60–65 per cent fall in crash injuries, yet many car companies failed to fit them until forced to much later by legislation. Even then, many drivers refused to wear them. Safety didn't become a big issue in most places until America's Ralph Nader hit the headlines with his controversial but hugely important 1965 book, *Unsafe At Any Speed*. 'Probably no other industry in this country [the USA]', he claimed, 'devotes so few of its resources to innovation of its basic product.'

The Standard Vanguard was one of several early attempts to give US big-car styling to little English cars. It was fine as long as you didn't look at it.

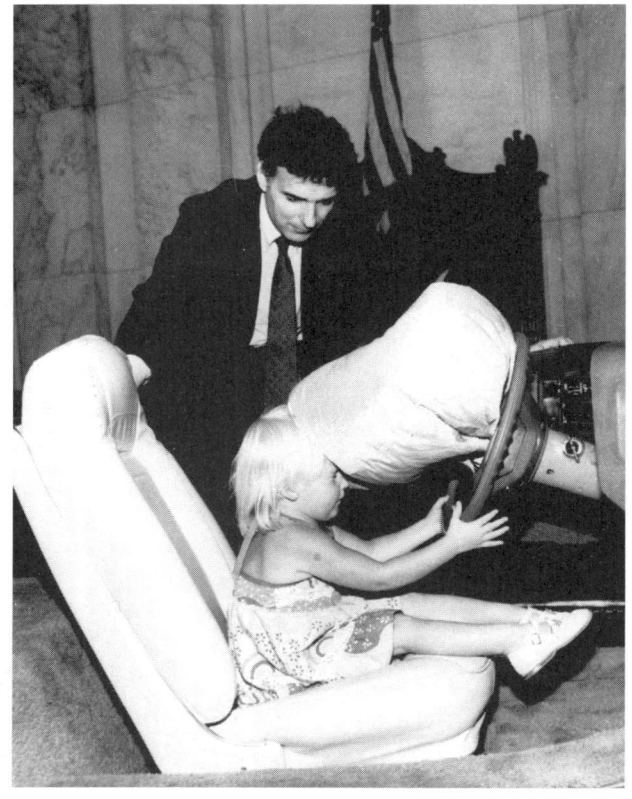

Consumer advocate Ralph Nader became one of the most influential men in the automotive industry despite only ever owning one car: a 1949 Studebaker. He fiercely launched into US cars of the 1960s, saying they were 'horsepowered beyond any level of sanity' and that 'auto companies subordinated engineering integrity to stylistic pornography'. Nader was a big advocate of airbags, despite the visibility problems they could cause young drivers.

At the time Nader was working on his book, more than 40,000 Americans were dying each year on the roads. Some southern European countries fared even worse in the same era, with a road toll twice as high as the US per kilometre travelled. The northern Europeans, however, showed far more concern about road safety and there was an almost unique preparedness of buyers in that part of the world to pay extra for safe cars. Volvo and Mercedes-Benz first catered for this in their home markets and then succeeded in making safety a selling point around the world. From the 1950s Mercedes-Benz began serious crash testing to develop 'safety cells' (rigid frames around the occupants) and 'crumple zones' (areas around the safety cell designed to crush in a controlled manner to lessen the deceleration forces in a collision).

The gulf between the European and the US approach continued through the 1960s and 1970s but, through it all, people everywhere were buying cars at levels previously unimagined. The total world production of automobiles was 8 million in 1950 but hit 12.8 million by 1960 and 20 million just six years later. No less than 30 million cars were produced in 1973, but before that year was out the boom was curtailed by the first oil shock.

The 'shock' was the result of Arab oil-producing nations

cutting production to force up prices, and withholding supply from countries that had supported Israel during the Yom Kippur War. There was spiralling inflation and widespread petrol shortages in the West, and suddenly a new automotive focus on economy. It provided an ideal opening for companies used to building cars for a small, resource-poor nation: Japan.

The VW Beetle still offered millions of Americans fuel-efficient motoring and smaller US cars were on the way, but in the meantime Datsun and Toyota could offer VW-style economy in much more modern and roomy designs. There was even a choice of body styles.

The story was the same in Australia, where the Japanese

Despite its humble origins, the Mini came to symbolise Swinging London of the 1960s and was popular with the rich and famous too. This is racing driver Jean Denton's paisley fabric-covered Mini. 'You can simply dust it clean with a clothes brush,' she explained.

Tough times in the UK were underlined with such ridiculous economy cars as the Winn electric. Despite the inventor demonstrating the manoeuvrability that could be achieved by using shopping trolley wheels, the English market remained in a no-Winn situation.

makers suddenly increased their market share dramatically, and put enormous pressure on the local divisions of the US and British car manufacturers.

Many Westerners accused Japanese car makers of being blatant copiers, an accusation that was then hard to deny, and which still hasn't been completely shaken. Yet even if the Japanese were not being overly innovative with the cars themselves, they were quietly revolutionising the way they were being manufactured. Working smarter (out of sheer necessity) and pioneering such concepts as just-in-time

> Morgans look like old cars, but they aren't replicas. In 2006 production at Malvern, England, was rattling along at 11 units a week with the wooden-framed Morgan 4/4, first seen in 1935, still on sale. The company claims this as 'the world's longest continuous run of any production car in the world'.

delivery, lean production and statistically-based continuous improvement, the Japanese began making cars that were often cheaper, better built, more reliable and more durable than those from European and American makers. Those who bought Japanese cars for their low price and fuel efficiency during the oil crises (there was a second shock in 1979, provoked by the Iranian revolution) discovered many other virtues.

That modern cars of all origins are so well built and well equipped, and are developed in such short periods and with so many variants, owes a great deal to the Japanese. The technical sophistication that Toyota, Nissan/Datsun, Mazda et al helped bring to affordable cars is also one of the

'Breaking in: How far you can break your Datsun Fair Lady will determine the amount of pleasure and advantages you can have with your Fair Lady'.

Datsun owner's manual (1963). Bizarrely, the little Datsun roadster was named after the musical *My Fair Lady*, courtesy of an Anglophile company president.

Just as daft, though a little more successful due to their government-subsidised use by invalids, were English three-wheelers. This is a Reliant 3/30 saloon. Despite the fins, it was rarely mistaken for a Cadillac.

Italian solution: the Fiat 600 in standard, convertible and 'Multipla' version, the last-mentioned possibly the smallest people-mover of all. Although Fiat reliability has led to the cruel suggestion by Anglophones that the name stands for 'Fix it Again, Tony', the Germans insist it's an acronym of *Für Italien Ausreichende Technik* ('sufficient technology for Italy').

'Would you buy a used car from this man?'

This phrase, summing up the ultimate criterion for trust, has been most associated with US politician Richard Milhous Nixon. Its first use in relation to Nixon (in the early 1950s) has been credited variously to comedian Mort Sahl, Nixon's political opponent Helen Gahagan Douglas and others.

In 1960 Citroën went wild and extended the 2CV colour choice beyond grey. However, even owners of these later and more sophisticated 2CVs often needed to ask a woman to lean against their car to stop it falling over.

reasons that the most modest of today's cars have several computer 'brains' to regulate major functions.

The microchip first gave us intermittent windscreen wipers and lights that dimmed automatically (come to think of it, Lucas gave us those features decades earlier, though not deliberately). Now the electronic brains installed in cars can think out the most efficient gear changes or the best way to atomise fuel during the combustion process to make engines run cleaner and more efficiently than was ever thought possible. Mercedes not only popularised antilock

> After World War II there was a flurry of patents for inflatable safety devices. Airbags as we now know them were first fitted to production cars during the 1970s; the 1974 model Oldsmobile Toronado was the pioneer, though interest was low. It wasn't until the late 1980s that they became mainstream items.

The VW heritage of the early Porsches – this is a 356 – was impossible to miss. The successor was to be known as the 901 model, until it was discovered that Peugeot had registered all the three-number automotive model designations with a zero in the middle. It became the 911. There was pressure to change the name after the events of '9/11' (September 11, 2001), but Porsche resisted.

brakes, it went further in the late 1990s with ESP (electronic stability program) systems that allowed the chip to take over crucial decisions such as braking and steering. Even some economy cars now have this feature.

It is a safe bet that cars will become markedly smarter, safer and more environmentally conscious with each year. They are also likely to become cheaper and more numerous as new low-cost makers enter world markets: Japan was followed by Korea; China looks likely to be the next major force.

In 1970 the Tass news agency announced the millionth Moskvich. That Russian people were prepared to wait years to secure one said more about the system than the car. They lined up to buy bread and toilet paper too.

THE BOOM TIME

The press release for the 1965 Aston-Martin DB6 said the car had a 'special raised lip at the rear to keep the car stable and firmly planted on the ground at high speed'. By later standards it was a remarkably modest 'airdam' (or 'wing' or 'spoiler'). Unfortunately the press release didn't explain how Valerie Leon, 21, of St John's Wood, managed to fit in with that hat on.

What hasn't changed – bizarrely – is that the reciprocating piston engine, as found in the 1886 Daimler and Benz vehicles, still powers most of the 60 million-plus new cars and light trucks being produced each year, along with the vast majority of the 500 million motor vehicles now crowding the world's roads.

The Mustang was Ford's super-success of the 1960s. Best of all, most buyers spent another 25 per cent of the purchase price on accessories.

141

Nissan built Austins after World War II, and not particularly well. Curiously, even Japanese cars sold within Japan have almost always had their badges written in English script, a tradition that probably owes itself to an early desire to make 'cheap' Japanese cars look like prestigious imports.

Toyota's Model AA from 1936 was clumsy and no real sign of what was to come.

The Rise of Japan

When America's car-making industry was first getting into its stride, Japan was a feudal, rural economy. After World War II, when Detroit was a boom town filled with confidence and unlimited promise, Japanese industry lay in ruins and what was left of management and labour were at war with each other.

The first private cars in Japan after World War II were not available until 1947, and only 110 were sold. Sales increased to 1594 three years later. Various Japanese makers produced often appalling replicas of European cars, while cork-maker Toyo Kogyo made three-wheeled trucks, choosing to name them after the Zoroastrian god of light, Mazda. Soichiro Honda took a different route, bolting war surplus power generators onto bicycles.

During the 1950s Japan's economy benefited enormously

from US attempts to create a bulwark against communism, but in the late 1950s all the Japanese makers combined still did not produce as many cars per year as Holden. However, in 1960, 10,000 Japanese cars were exported and a miracle was clearly underway. Japanese cars became much more numerous during that decade, and aside from the crude three-wheelers still built for the local market, they were also becoming more sophisticated.

Mazda eventually moved from three-wheelers to four-wheelers. Here's how Sydney's *Sun* saw the pending arrival of what became known as 'the light bulb car'.

During the ascendancy to car-making superstardom, the Japanese makers refined almost everything to do with production engineering and manufacturing, and the faddish domestic market taught the makers to react quickly to customer demands. This domestic market was the

The rare-as Toyota 2000GT convertible. The large, offset bonnet mascot didn't make it into production.

> Interesting Japanese model names: Today Humming (Honda small car), Debonair Exceed Contega (Mitsubishi limousine), Mitsuoka Viewt (pseudo-Jaguar named by someone tired and emotional . . . as a viewt), Proceed Marvie (Mazda 4WD, not to be confused with that other Mazda 4WD, the Bongo Brawny) and Avenir Salut Aero Express (Nissan wagon).

government's gift to its car makers. It was kept almost completely free of imports, enabling Japanese car makers to charge more for their cars domestically than they did overseas, while stringent 'safety' laws forced consumers to scrap most new cars within four years. This hugely increased annual sales. The Japanese consumer was, in effect, subsidising the car-exporting miracle.

In 1980 the Japanese industry built 7 million road vehicles, surpassing the US industry for the first time. Ten years later a record 13.5 million were made in Japan, representing about 25 per cent of the world's production for the year.

Down, Rover

In 1961 car buyers could consider a Mini-Minor, a Jaguar E-Type, an Austin-Healey 3000, a Land Rover Series II or a Rolls-Royce Silver Cloud. All of them were world-beaters and all of them wholly British. But from there it was, alas, mostly downhill.

By the 1980s there was only one surviving British-owned mass-market brand: Rover. Like an automotive black hole, it had sucked in everything around it over the years. At its most expansive, the Rover Group (under a variety of names including BMC and British Leyland) controlled Albion, Alvis, Austin, Austin-Healey, B.S.A., Crossley, Daimler, Jaguar, Lancester, Land Rover, Leyland, Maudslay, MG, Morris, Riley,

> 'The Ford Cortina will benefit from modifications to meet Japan's unique safety and emission regulations — details of which were still uncertain at the end of January, despite the fact that they will be mandatory from April 1 this year.'
>
> An interesting paragraph from a 1975 British Industry Council press release. It amply illustrates how Japan's 'free' market, which isn't free now, wasn't free then either.

THE BOOM TIME

The long, unwinding road. Britain's Jaguar company started as SS Sidecars (making motorcycle sidecars), but graduated to four-wheelers before World War II. The E-Type Jaguar was probably its biggest success – fast, svelte and in huge demand. The shape was, as many observers noted, extremely phallic.

Rover, Standard, Thornycroft, Triumph, Wolseley and other brands besides.

Yet these marques withered or were sold, and the parent company was forced into an unequal partnership with Honda. From 1984 most of its cars were merely Anglicised, Rover-badged versions of Japanese models. Bizarrely, in 1994, BMW somehow arrived at the conclusion that the only thing wrong with the British motor industry was that it wasn't owned by Germany.

> During the 1960s BMC produced 60 different Mini-Minor speedometers to fulfil the requirements of all markets and models.

> In the mid-1980s, the Jaguar XJS, a premium-priced sports coupe, was rated as the single worst quality car in the USA.

> 'BMW has been damaged by arrogance and ignorance. The adventure could not have ended any worse than it did.'
>
> Daniela Bergholdt, a lawyer for a German shareholder group, comments on BMW's purchase and eventual sale of the British car maker Rover (2000). Bergholdt wasn't entirely correct. BMW could have revived the Triumph name, something it was at one time seriously considering.

BMW purchased Rover outright and immediately started paying through the nose for the decision. The German press started referring to the company as 'The English Patient' and, by the time BMW finally ridded itself of Rover in 2000, the total losses were believed to total US$6 billion. Although BMW managed to sell the Land Rover arm to Ford for a decent sum, it could realise just £10 for Rover's passenger car operations. As part of the extraordinarily one-sided deal, BMW also had to lend the buyer, a group called Phoenix Consortium, $1.35 billion and give it 80,000 unsold cars (worth $2.7 billion) just to take the mess off its hands.

Some briefly rejoiced that Rover was British again, but by 2005 it was clear that, after decades of defying gravity, the withered rump of the British car industry was finally about to start pushing up the daisies. What was once one of the

The other end of the scale: the Triumph Herald was gruesomely unreliable and if you listened carefully you could hear it rust.

What about all those other famous British brands that didn't end up as part of Rover . . . brands such as Humber, Sunbeam, Vulcan, Singer, Hillman and Talbot? They mostly went to Chrysler Europe in the late 1960s when it bought the Rootes Group. Here a Humber Hawk Saloon is seen jumping with joy at the prospect. In 1978 Peugeot procured the remnants of Chrysler Europe for the grand total of $1.

world's biggest manufacturing conglomerates collapsed with an ignominy that was entirely consistent with its latter years. Huge government (and BMW) grants had been swallowed, several executives had made themselves very wealthy while filling fields with cars that nobody wanted to buy and thousands of workers were thrown onto the streets.

The rights to produce the Rover 75, the only slightly modern car in the line-up, somehow ended up in the hands of the Shanghai Automotive Industry Corp, while the once-revered MG brand had been so comprehensively run down that no mainstream buyer could be tempted to show even the slightest interest.

Those major British brands that have made it into the twenty-first century have done so with foreign ownership. They include Jaguar (owned by Ford), Bentley (VW), Land Rover (Ford), Rolls-Royce (BMW), Aston-Martin (Ford) and Vauxhall (a much-earlier General Motors acquisition). Out of the wreckage of the Rover debacle, BMW had one win: it relaunched the Mini as a stand-alone brand and kept it for itself.

'There is no way we will start providing any sort of assistance unless it helps the situation.'

British Prime Minister Tony Blair, speaking in 1999, fails to throw much light on his government's policy toward the troubled Rover Group. The concern known for years as 'the troubled Rover Group' finally became 'the failed Rover Group' in 2005.

STEP ON IT!

Go configure

When Volkswagen launched a V5 engine in 1997 it was unusual enough for people to ask whether there was a typo on the boot-lid, or whether the cylinders had been counted out by the bloke responsible for the nuts, bolts and allen keys at IKEA. The VW salesfolk, however, assured potential buyers that it was definitely a V5, and that no cosmic or even national laws had been broken by producing a vee engine with an uneven number of cylinders.

In the past we've seen many vee engines: the V2 (Morgan and others), the V4 (Ford and more), the V6 (almost everyone), the V8 (ditto), the V10 (Chrysler Viper and BMW M5), the V12 (in sports cars such as Ferraris as well as limousines) and the V16 (Cadillac, Marmon and a couple of others). Born-again Bugatti has the Veyron coupe with sixteen cylinders arranged in a double-vee (or W) pattern.

Henry Ford's X engine, with eight cylinders. He wanted to use it for the Model A and it was only the brave decision of Ford's engineering chief to stand up to him that kept this noisy, unreliable donk out of production. The prototype ended up powering Henry's personal motor launch.

However, the V5 remains the only 'uneven' vee, horizontally opposed or double-u engine to go into volume production.

If you count inline engines as well, there aren't many slots unaccounted for between one and twelve (and let's face it, having more than 12 cylinders is just being silly). One-cylinder automotive engines were once common, while three-cylinder donks have been

---

In March 1966, GM Corporation built its 100 millionth car, the first company to do so.

Bugatti's EB 18/3 Chiron was the product of the reborn, VW-owned company. The 6.3-litre engine had 18 cylinders. Production did not eventuate, the reborn company opting for a modest 16 cylinders in the first model to go on sale, the 2005 Veyron.

built by Daihatsu and others. Audi marketed the first five-cylinder petrol engine cars in 1977, though Mercedes-Benz had already offered a diesel in this seemingly odd configuration. Several companies have built straight eights, while Willis and Maywood, an Illinois concern of the late 1920s, powered a car and truck with a radical, though not radically successful, 'straight nine'. By one report Spain's National Pescara company built a straight ten in the early 1930s. This wasn't the longest inline automotive engine, though. French maker Voisin built a straight twelve prototype. On this basis, seven and eleven may be the only cylinder counts missing from the automotive kingdom.

> 'The best way to fight Communism in this country is to give each worker a scooter, so he will have his own transportation, have something valuable of his own, and have a stake in the principle of private property.'
>
> Motor scooter manufacturer Enrico Piaggio explains to *Time* magazine (1952) how his Vespa two-wheeler will keep Italy free of Communism. Perhaps this is the origin of the phrase 'the workers have nothing to lose but their chains'.

In 1948 there were 1 million vehicles on Australian roads; seven years later that number had doubled. The 3 million mark was passed in 1962 and the progression has continued: Australia will soon acquire its 15 millionth registered road vehicle. That is three vehicles for every four persons, and the number of vehicles almost certainly now exceeds the number of people with licences.

CHAPTER NINE

Off the track

> 'If a new Jeep vehicle can't take you there, maybe you ought to think twice about going.'
>
> Snappy Jeep slogan from 1972.

When General George C. Marshall referred to 'America's greatest contribution to modern warfare', he wasn't talking about the nuclear bomb, napalm or euphemisms such as 'collateral damage'. He was talking about the humble Jeep.

About 645,000 examples of this pioneering four-wheel drive, go anywhere, do-anything vehicle were built between 1940 and 1945 by Willys, Bantam and Ford. After VJ day, tens of thousands of them were left scattered all over Europe, Asia and Australia in various states of repair. Despite the occasional bullet hole or body part wedged under a wheel arch, they were snapped up by farmers, ex-soldiers, home mechanics and adventurers and a new pastime was born: recreational four-wheel driving.

This was a huge surprise to Willys-Overland Motors, the company that had built the greatest number of Jeeps during the war and which trademarked the Jeep name when it was finished. The company had thought that if it could get away

The Jeep: a rough but effective war baby.

with building such an unrefined and unstylish vehicle for anyone other than the army, it would be solely for farmers. Little did it realise that the main use for 4WDs would eventually be commuting around cities, and that only about 1 in 20 would ever be taken off road.

The first truly successful 4WD to be born in peacetime came not from the USA but from Great Britain. It owed a lot to the Jeep, however, and a lot to the war. It was called the Land Rover and went on to be perhaps the most important Rover of all, arguably the one that saved the famous English company – for a while – in the dire days after World War II.

The Rover company had started in 1904 and was in pretty serious trouble by the end of the 1920s. It recovered

In civvies: despite the ever-so-slightly more stylish bodywork, the Jeepster Sports Phaeton (1948–1950) was still army issue underneath.

England's Jeep, the Land Rover, proved even more versatile than the original and was soon to be found working for its living all over the Australian Outback.

and spent the late 1930s and early 1940s making aircraft engines and airframes. Towards the end of the war, civilian production resumed but Rover was soon in dire financial straits again. It needed a new vehicle it could develop quickly, with minimal cost, and ship offshore to meet a new government directive on exporting (basically: bring in foreign capital or you won't be allocated the materials needed to meet local demand).

The Land Rover design brief, legend has it, was 'make a Jeep using as many existing Rover components as possible'. Work began in early 1947. The first prototype – sitting on a Jeep chassis frame and containing many other Jeep components – was soon finished. It had a central driving

Or in motor sport – either as a competitor or rescue vehicle.

Even companies such as Alfa Romeo got into the four-wheel drive act, believing there would be a huge agricultural market for such vehicles. This, the Alfa Matta, was not as pedestrian friendly as it might have been.

Citroën took on the Himalayas, Africa and other challenges before World War II with half-tracked vehicles. This 1931 expedition went from Beirut to Beijing, covering 12,115 kilometres over 315 days. The vehicles had to be dismantled to make it through some of the narrower Himalayan passes.

The Lunar-Rover, the 'moon car' first used in 1971, covered 96 km and hit a top speed of just under 17 km/h on the lunar surface during the Apollo 16 mission. Four examples were built at an estimated cost of US$4.75 million apiece. Three went to the moon.

The M-274 or Mechanical Mule was an attempt by Willys to out-Jeep the Jeep. It was available with two- or four-wheel drive and two- or four-wheel steering and the controls were designed so the Mule could be operated even if the 'driver' was crawling beside it.

position to get around the old left- or right-hand drive dilemma, and was so sparse doors were considered an unnecessary luxury. The body was aluminium because steel was harder to come by.

The Land Rover Series 1 went on sale in 1948 and was an instant hit. For farmers, it was more versatile than a Jeep because it had power take-offs and could be bought with a variety of body styles. It was also a success with construction workers and those who wanted to head for the remotest corners of the earth. Land Rovers soon became a common sight in Australia and most other places with a lot of unsealed roads.

General Electric, meanwhile, seriously experimented with a vehicle that walked over objects. The US Army stumped up with research money, presumably fearing that the Russians were working on a vehicle that could jog or even sprint.

The 1960s gave us hover-cars and other go-anywhere designs said to be able to traverse water, mud, sand and rocks, but none succeeded as personal transport. Lockheed's attempt was the TerraStar, with clusters of three wheels on each corner enabling it to 'walk' up and down stairs and create three times as much repeat business for tyre suppliers.

Willys, meanwhile, branched out with civilian 4WDs for commercial and leisure use. It wasn't until 1962, though, that we saw the first Jeep model not based on the military design.

The success of Jeep and Rover during the 1950s also dragged in the Japanese. The vehicle that would become the Toyota LandCruiser was first seen in 1951 and known as the Toyota Jeep. Legal problems with the name followed. Mass-production started in 1953 but the vehicle was still a me-too version of the US Army Jeep, powered by a pre-war Toyota truck engine. Unexpectedly successful, it saved the company from bankruptcy.

The name LandCruiser was adopted in 1954, although Toyota sometimes spelt it as one word and sometimes as two, or as a compound word with a capital C in the middle. The early Toyota passenger cars offered in the US were so unlovable they were withdrawn from the market during the

Dune buggies were a big vogue in the 1960s. Most were modified and very second-hand VWs.

'Auto Industry Agrees To Install Brakes In SUVs.'

Headline from American satirical journal *The Onion* (July 2000). SUVs are Sports Utility Vehicles – the American name for large 4WD wagons.

early 1960s. In contrast, the LandCruiser was well-received and kept the brand name alive Stateside until the launch of the 'new generation' Corona in 1965. The LandCruiser would eventually outsell the Jeep and Land Rover in most markets.

Another big Japanese player arrived almost by accident in the early 1970s when a Subaru engineer answered a question no-one except a few Japanese farmers were asking. He produced a low-cost, all-wheel drive version of the front-

Eventually buggies became big business, gaining sleekish fibreglass bodies along the way. This is the most famous variant: the much-copied Meyers Manx, built on a shortened Beetle platform.

drive Subaru wagon. Sure, it wasn't the world's most capable off-roader, but it could go some of the places a full-size four-wheel drive could go and was a great deal cheaper to buy and run. The 'Subie' wagon became a big hit with Japanese farmers, before its success spread around the world.

At the other end of the scale, Rover realised that the civilian market for 4WDs wasn't restricted to low-budget adventurers or people with straw hanging from the sides of their mouths. The landed gentry could easily be convinced they needed a vehicle that looked the part, even if the driveway to the

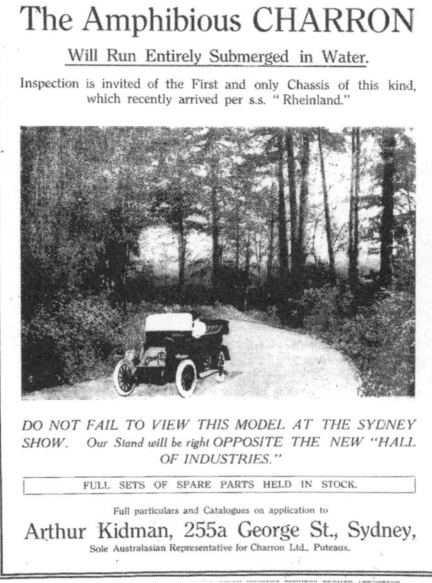

The ultimate off-road machine should be able to go off-shore as well. This Charron was apparently designed not to float on the water but to drive under it. Shame about the open cockpit then.

Unfortunately, most car-boat hybrids proved to be bad cars and even worse boats. This Boatswagen, a modified VW photographed on Sydney Harbour in the 1960s, looks to be no exception, struggling in calm water.

> According to US statistics published in the 1990s, a fatal car accident occurs every 5500 vehicle years, an airbag deploys every 175 vehicle years, a 'fender bender' occurs every 5.5 vehicle years and a panic stop utilising antilock brakes occurs once each vehicle year.

The Range Rover was ideal for roadside dental inspections, while the special design of the two-piece tailgate enabled objects to be packed and unpacked while smoking.

country manor was fully sealed. The result was the Range Rover.

While Jeep had already enjoyed some success with big 4WD wagons in the States, they were truck-like. The 1970 Range Rover had a big V8 engine, luxury trimmings, a versatile (perhaps even classic) body-style and a suspension and braking system far more refined than anything previously

The race to build Australia's first mainstream off-roader could have been won in 1972 if Ford had put more support behind this. And if it had been something other than a dreadfully bad vehicle. It comprised a Falcon ute body crudely dumped on top of a 4WD Jeep platform. Ford made about 430 examples.

found on a 4WD. The luxury 4WD was born and now almost every prestige manufacturer has a luxury 'off-roader', designed first and foremost for the bitumen, as well as a range of cheaper 'family 4WDs', also primarily for the blacktop.

In one of the truly bizarre twists in motoring history, the freedom and adventure machine, made big and tough for extreme conditions, had become instead the armoured cacoon for anxious urbanites.

The First Gulf War (1990-1991) led to an even more extreme phenomenon: public affection for the Hum-Vee four-wheel drive. This had been built for the US Army by AM General Corporation, the military division of Jeep, which had been spun off separately in 1971. A civilian version known as the Hummer (the soldiers' nickname for the military version) was quickly produced and sales were surprisingly strong despite it weighing nearly 3.3 tonnes. In 1999 GM bought the rights to market a full line-up of Hummer vehicles, ranging from enormous to gargantuan. The 'Jeep' cycle had started all over again.

> 'The car has become the carapace, the protective and aggressive shell, of urban and suburban man.'
>
> Marshall McLuhan, Canadian social commentator, in his book *Understanding Media* (1964).

The civilian version of the military Hummer became a big success despite being ridiculously overweight, socially irresponsible and totally inappropriate for road use. Or perhaps because of it. Bodybuilder-turned-actor Arnold Schwarzenegger (later to act as governor of California) is a big fan.

> **The Names of the Beasts:** Cars named after mammals include Chevrolet Impala, Triumph Stag, Ford Mustang, Ssangyong Musso (it means rhinoceros), Audi Fox, Mitsubishi Colt, Singer Gazelle, Fiat Panda, Renault Dauphin and Topolino (Italian for little mouse and noisy Fiat).

Just a Jeepster

Lesser known Jeep milestones:

- In 1938 the US Army began the search for a car to replace the motorcycle–sidecar combos it used for messenger and reconnaissance duties. One concept considered was the 'Howie Bellyflopper', a small four-wheeler requiring the driver to lie on his stomach to keep the vehicle's profile low.

- Eugene the Jeep, a small animal that could solve problems and 'travel back and forth between dimensions', appeared in a 1936 edition of E.C. Segar's *Popeye* comic strip. It is widely believed that this provided the origin of the name 'Jeep' for the US Army vehicle, although some claim it was a contraction of GP (for General Purpose), and others point out that the term 'jeep' was also used in World War I for an unproven recruit or machine.

- In 1948 the two-door 'Jeepster' model was launched. In the early 1970s glam-rock band T. Rex had a enormous hit with a song entitled *Jeepster*. In the chorus lead singer Marc Bolan, for no known reason, declared to a girl that he was just a Jeepster for her love.

- In 1953 Willys-Overland was bought by the Kaiser-Frazer group, in 1970 Jeep became part of American Motors Corporation and in 1987 Chrysler captured the famous marque (Jeep is still a division of Chrysler, which in turn is now part of DaimlerChrysler).

The 1940 Quad – the Willys proposal for the US Army. Some Quad design aspects were taken into the eventual vehicle, the Jeep. It was also pug ugly, for example.

Volkswagen's Schwimmwagen was based on Germany's World War II Jeep, the VW Kübelwagen, or 'bucket car'. It had four-wheel drive on land, and a propeller to push it along at about 10 km/h in water. There was a snorkel exhaust, even a paddle if the dak-dak plop-plopped. Surviving examples are few: they were built primarily for the German Army's Russian campaign and this proved to be a test so tough not even VWs could stand it. Just 14,276 examples of the drip-dry VeeDub were built.

> While being driven through the streets of Dallas, Texas, in a Lincoln Continental, John F. Kennedy was assassinated by somewhere between one gunman (Warren Commission) and several dozen (Oliver Stone). Surprisingly, the car in which the President died wasn't scrapped. Instead it was further customised, receiving a roof and more bulletproofing. It continued to be used by US Presidents until well into the 1970s. It is now on display at Detroit's Henry Ford Museum near the rocking chair in which Abraham Lincoln was shot, one of the six almost-priceless Bugatti Royale limousines and a test-tube said to contain Thomas Edison's last breath.

Going for a song

Jackie Brenston's 'Rocket 88', the 1951 song considered by many to be the first ever rock-and-roll track, concerned an Oldsmobile. Since then, popular singers have sung of Chevys (beside levees and elsewhere), Ford Thunderbirds and, arguably more than anything, Cadillacs. Even British bands raised on Austin and Morris models have recorded songs about that befinned piece of aspirational Americana.

Marc Bolan, singer and guitarist for T. Rex, and serial wearer of sparkly clothes. He sang of Jeepsters and in 1977 died while passenger in a Mini-Minor.

Car songs actually go back at least to 1905, when Gus Edwards and Vincent Ryan wrote '(Come Away with me Lucille) In My Merry Oldsmobile' and the music halls rang out with such ditties as 'He'll Have To Get Under, Get Out And Get Under (To Fix Up His Automobile)'.

The 1950s brought to prominence the man who was probably the greatest ever singer of songs about cars: American Chuck Berry. His songs included the tale of the 'Jaguar and the Thunderbird', racing for the county line with the police car one yard behind, plus 'No Money Down', 'You Can't Catch Me', 'My Mustang Ford' and many more. From the early 1960s The Beach Boys

---

Ten songs dealing with specific cars: 'Little GTO' (Ronnie and the Daytonas); '409' (The Beach Boys); 'Ford Mustang' (Serge Gainsbourg); '455 SD' (Radio Birdman); 'Little Red Corvette' (Prince); '2CV' (Lloyd Cole and the Commotions); 'Brand New Cadillac' (The Clash); 'Hot Rod Lincoln' (Johnny Bond); 'Mercury Blues' (Ry Cooder); 'Grey Cortina' (Tom Robinson Band).

> Keith Moon, the eccentric drummer for the English band The Who, supposedly almost killed himself driving into a swimming pool in that most un-Continental of cars, the Lincoln Continental. Although the story is almost certainly apocryphal it has become an iconic symbol of excessive rock star behaviour. British band Oasis paid tribute by depicting a Rolls-Royce Silver Shadow in a swimming pool on the cover of its album *Be Here Now*.

took up where Chuck Berry left off (he was in jail, actually, on dubious charges of violating the even more dubious *Mann Act*), producing a swag of car-related songs for a newly motorised American youth. There was 'I Get Around', 'Shut Down', 'Little Deuce Coupe' and 'Fun, Fun Fun', with its much quoted line about daddy taking the T-Bird away. The late 1950s and early 1960s also brought a spate of teen-death-car-crash-horror songs of the ilk of 'Tell Laura I Love Her'.

The Beatles (shown below in Adelaide, 1964), rarely sang about matters automotive and no car is mentioned by brand in any Lennon–McCartney song. However, there was a VW Beetle on the cover of the *Abbey Road* album. The actual car – numberplate LMW 28IF – was auctioned by Sothebys in London in 1986 but fetched just £2300 pounds (about AU$5000).

STEP ON IT!

> The first car owned by rock 'n' roll singer Chuck Berry was a 1934 Ford V8 that cost him $35. The second was a 1937 Oldsmobile and using this car he and some friends committed a series of armed hold-ups in 1944. This landed Chuck a ten-year jail sentence but he was released early, hit the big time, bought a pink Cadillac and sang about this and many other cars besides.

What followed from the late 1960s were heavier, roaring down the highway numbers such as Steppenwolf's 'Born To Be Wild'. Bruce Springsteen created a neon-lit motor world on the 1975 album *Born to Run*, with a barefoot girl sitting on a Dodge and a hard-up Lothario offering his catch of the day redemption from beneath his dirty hood. Perhaps you had to be there to get the measure of it, but millions of record buyers took the ride in 1975 and both *Time* and *Newsweek* magazines put 'Loose Windscreen' on the cover.

As the 1970s became the 1980s, Marianne Faithful sang of 'Lucy Jordan' and her ambition to drive through Paris in a sports car with the warm wind in her hair. But the new musical decade was less car-oriented, and the nineties and noughties provided leaner pickings still. Not only was youth

From the early 1960s, Ed 'Big Daddy' Roth developed a fusion of pop art, automotive culture and wild cartooning. His character Rat Fink (a drooling, fly-ridden rodent who drove monstrously modified cars with huge tyres and engines) influenced Mambo and other modern designers. Roth was described as the 'Dali of kar kulture' in Tom Wolfe's book *The Kandy-Kolored Tangerine-Flake Streamline Baby* but after the oil shock of the early 1970s he fell from favour. Roth's unlikely cover artwork for the 1982 album Junkyard, by the Australian band The Birthday Party, either provoked or coincided with a strong Rat Fink revival.

more excited by things other than cars, but songwriters no longer felt the need to bother with innocent metaphors for sex and drugs. This robbed the car of one of its most potent artistic uses. Automotive images did still pop up occasionally (the Prodigy's 'Diesel Power' is one example, Radiohead's 'Airbag/How Am I Driving' is another, while rap songs regularly name-check favoured Blingmobiles using such abbreviations as Cad, Lac, Benzo, Lex and L-dog, for Lincoln), but the old days of singing about a diff ratio or brand of carburettor seem well behind us.

> 'If the fluctuations of the market are erratic, there is one safe bet in owning old cars: the re-builds will rarely, if ever, come in below budget or on time.'
>
> Restoration advice from Pink Floyd drummer and inveterate car collector Nick Mason (1998).

> 'There was nothing left of it, except for the tailfins.'
>
> A witness describing the red Plymouth that Australian rock singer Johnny O'Keefe drove at high speed into a gravel truck (1960).

The Aerocar wasn't as long as the trailer that carried its wings. Still, the Aerocar really did fly, though not out of showrooms. Production ran to six.

James and Rolf

James Dean, a young American actor with just one released feature film to his name, effected one of history's better-known car crashes in a Porsche Spyder 550 on 30 September 1955. Dean's car was just two weeks old and he was driving from Los Angeles to compete in a car race at Salinas Municipal Airport.

Dean had formerly owned a Porsche Speedster and had graduated to the Spyder 550 as soon as his income permitted it. Cheap it wasn't – it cost $7000 in the days when a Cadillac could be had for under $4000 – yet Dean immediately dubbed it 'The Little Bastard', had the number '130' painted on its sides and tail, and planned its first race.

But on the way Dean managed to kill himself after colliding with a Custom Deluxe Club Coupe Sedan at the junction of Highway 466 and 41, near Cholame, California (USA).

Less well known is that at the time of the accident Dean had a passenger. Dean had bought the Porsche on the proviso that the Beverly Hills Porsche dealership provided a mechanic to accompany him to the race. A German mechanic named Rolf Weutherich, who had been the 42nd Porsche employee, was the unlucky man supplied. Dean was killed instantly; Rolf survived but spent the next 18 months in hospital.

Ironically, Dean had just filmed a road safety commercial in which he said: 'Take it easy driving – the life you save may be mine'. The driver of the Ford with which Dean collided was widely reported at the time to be Donald Gene Turnupspeed,

> **'That guy up there's gotta stop. He'll see us.'**
>
> The last words of James Dean (1931-1955), as reported in many different books and magazines. The mystery is who heard them uttered. In a rare interview many years later, his only passenger, Rolf Weutherich, said he was asleep when the accident happened.

In the French language a car is feminine but a truck is masculine. It does not follow, however, that all parts of a car are feminine or all parts of a truck are masculine.

James Dean and the Porsche Spyder 550 he bought to race. There was no Hollywood ending.

which seemed to add a further creepiness to the whole incident. In fact his name was Turnupseed, which, as names go, is not much less creepy.

The crash was the catalyst for all sorts of stupid carryings-on, including an Annual James Dean Memorial Car Rally retracing the route Dean and Weutherich took. There are many websites on the subject of the accident and a model-maker named Jeff Hodgson builds scale replicas of the smash site, with crumpled cars in situ. And considering that the time between Dean losing control and dying has been estimated to be less than 10 seconds, it may seem a bit rich to string it out to 200 pages, yet American author Warren Newton Beath — seemingly totally preoccupied with the length of the skid marks, angles of deflection and so on — does so in his book *The Death of James Dean*. In fairness, however, he also includes a short history of the car after the accident.

According to Newton Beath, a little over a year after Dean's death, an orthopaedic surgeon named Troy McHenry was racing a Porsche Spyder 550 at a Los Angeles racetrack when he lost control and smashed into a tree. He died in the wreckage. Unbeknown to him, some of the suspension parts in his car were from Dean's 550, having been taken from the wreckage and sold by the insurance company. The stripped body of Dean's car was turned into a travelling safety exhibit. In 1959, again according to Newton Beath, the truck carrying it was involved in an accident and the Porsche slid forward and crushed the driver, George Barhus, to death.

Meanwhile, Rolf Weutherich successfully rallied with Porsche in the 1960s (and navigated for the team that came second in the Monte Carlo Rally in 1965, though his name appears in the official results with slightly different spelling). He left the Porsche company in 1968 and died in the wreckage of a Honda in 1981.

> 'The car is a big powerful weapon. It's exactly the same as a gun. If you put a gun in somebody's hands you are going to change their personality, you are going to affect what they are capable of. With a car it is exactly the same.'
>
> British psychologist Conrad King (1998).

Honda's first prototype Formula One car was called the RA270. The numbers came from the engineer's objective of reaching a top speed of 270 km/h. Projected top speeds (in miles per hour) also featured in the names of the Jaguar XK100 and XJ220.

CHAPTER TEN

Addicted to speed

While road racing, then circuit racing, was capturing the popular imagination at the dawn of the twentieth century, a completely different form of motor sport was being carried on by a few dedicated, often overlooked, and usually completely mad adventurers. Their quest was the land speed record (LSR).

Trying to be the fastest man on earth was a dangerous business. Early cars could do evil things at even modest speeds, and were often deadly at the limit. The science of aerodynamics, for example, was somewhere between rudimentary and non-existent. As speed mounted, cars were likely to suddenly lose bodywork, tyres or their steering capability.

Despite this, the LSR chase started early, with Gaston de Chasseloup-Laubat's electric-powered *Jeantaud* setting a mark of 63.2 km/h in 1898. A year later a Belgian, Camille Jenatzy, used a cigar-shaped electric car to beat the

> 'Before long my husband may be the fastest man on earth, or I may be a widow.'
>
> Tonia Bern, wife of Donald Campbell, speaking in 1963. By 1964 he was; in 1967 she was.

This battery-powered cigar on wheels was used to set early land speed records. It was known as *La Jamais Content*, or 'The Never Happy', though 'The Never Even Slightly Safe' might have been more accurate.

Frenchman's speed and break the 100 km/h barrier for the first time. Henry Ford drove a self-built car at 147.1 km/h in 1903 and a year later the 150 km/h, then the 100 mph (161 km/h) marks were exceeded. In the USA in 1906, Fred Marriott drove a steam-powered Stanley Steamer Rocket at 205.3 km/h.

From the 1920s a string of Brits – Kenelm Lee Guinness, Malcolm Campbell and Henry Segrave – started to dominate. They pushed the LSR past 300 km/h by 1927 and, in September 1935, on salt flats in Utah, Malcolm Campbell drove his famous aero-engined *Bluebird* at an extraordinary 301.13 mph (484.60 km/h).

When World War II broke out, the record stood at 595.3 km/h, set by British fur-trader John Cobb in his *Railton Mobil Special*. Cobb would return in 1947 to beat his own record with 634.27 km/h. This would stand until the

The 999 racing car – more an engine on rails – won the 5-mile Manufacturers' Challenge Cup of 1902 in the hands of former boxer and cyclist Barney Oldfield. He is seen at the wheel, or tiller, next to the car's builder, one Henry Ford.

The Oldsmobile Pirate, built in 1903, consisted of little more than a one-cylinder engine, four wheels and a seat. It completed 5 miles (about 8 km) in 6.5 minutes before being spruced up and renamed 'The Flyer'. In its new guise it completed a mile in 42 seconds.

1960s when a series of brave and often short-lived Americans used jet engines to power their machines (by this time Cobb had killed himself travelling in a boat at about 320 km/h at Loch Ness in Scotland).

This jet-powered trend horrified purists, while the unpredictable nature of these new machines caused many fatalities. Still, the 'land missiles' of Tom Green, Art Arfons and Craig Breedlove swapped the record eight times between

Nothing spindly about the Blitzen, or Lightning, Benz. Just four cylinders but 21.5 litres and a claimed 200 horsepower (150 kW). Not sure about the helmet being worn here by Bob Burman at the Indianapolis Brickyard though. In 1911 at Daytona Beach, Burman covered a mile at an average of 228.1 km/h with a flying start.

If Benz thought that 21.5 litres was pretty big for a four-cylinder engine, Italy's Fiat company was out to show the Germans it was not going to be intimidated. Its Type S76 from 1910 had a 28.34-litre 'four pot'. And who are you to say it looks ridiculous?

> 'An essentially 20th century art, and one demanding as much theoretical study, natural flair, learning and practice as any of the classical arts.'
>
> Legendary British motor sport writer Denis Jenkinson (1921-1997) on controlling a car at high speed. He believed it was as worthy a pursuit as music, painting and literature. From his book *The Racing Driver*.

1963 and 1965, raising the mark to just shy of 1000 km/h.

In between, Malcolm Campbell's son Donald travelled to Lake Eyre in South Australia with a completely new *Bluebird*. It was there that this odd, complex, superstitious and charismatic Englishman set his only world speed record on land, but even this was muddied by controversy. The 648.73 km/h he hit in the *Bluebird* in 1964 was faster than any other wheel-driven vehicle in history, but 7 km/h slower than Craig Breedlove had achieved with the jet-propelled *Spirit of America*.

During his 1964 Australian sojourn, Campbell also claimed the world water speed record, raising it to nearly 450 km/h on Lake Dumbleyung in Western Australia. Then, as on all other occasions, he avoided anything green, carried his Mr Whoppet teddy bear in the cockpit and wouldn't let anyone wish him good luck, because that was bad luck.

> The biggest-engined car of all time? The *White Triplex*, which broke the world speed record in 1928, achieving a bit over 334 km/h. It was powered by three Liberty V12 aero engines, giving it more than 81 litres capacity from 36 cylinders.

Malcolm Campbell's 1935 *Bluebird* on the way to a world record.

And his son Donald's more svelte 1960s *Bluebird*, standing perfectly still.

By many reports Donald Campbell was terrified doing what he did but, pushed along by the ghost of his late father (Sir Malcolm Campbell, the first man to drive a car at 300 mph), Donald took cars and boats to record speeds in the 1950s and 1960s. He fatally crashed his *Bluebird* boat at nearly 500 km/h on Lake Coniston, England, in January 1967.

Campbell argued that a jet car was not a car at all, and therefore shouldn't be officially recognised as a land speed record contender. Officials initially agreed, but soon jets, rockets and other propulsion systems were sanctioned and two LSR classes adopted. For outright honours, it was obvious that wheel-driven machines were no longer competitive.

America's Gary Gabelich attacked the LSR in 1970 in the three-wheeled *Blue Flame*, powered by a liquid natural gas-hydrogen peroxide rocket. This propelled Gabelich through the 1000 km/h barrier, and presumably lightened his hair colour at the same time.

The obvious next target was the speed of sound. In 1980 Stan Barrett used a Sidewinder missile to power his needle-thin *Budweiser Special*. He was clocked at 1028.1 km/h, but

Dr Nathan Ostich made the first jet-powered assault on the Land Speed Record with the *Flying Caduceus* in 1960. The caduceus is the legendary staff carried by Hermes, or Mercury, and is also an emblem of Ostich's medical profession. Although the initial diagnosis was positive, it hit only 355 miles per hour (571 km/h) and failed to wrest the LSR.

> The world Land Speed Record of 1227.952 km/h compares poorly with the fastest spaceships. America's Space Shuttle can accelerate from zero to 5533 km/h in two minutes flat, and orbit the Earth at about 28,200 km/h. The unmanned Helios 1 and Helios 2 probes hit 252,800 km/h while orbiting the sun.

his claim to have broken the sound barrier was widely disputed and, anyway, his run was in one direction only. (International rules called for a 1 mile 'pass' to be made in two directions within one hour to negate any wind or gradient advantages.)

In 1983 British businessman Richard Noble drove his *Thrust 2* jet car in two directions at an average of 1020.0 km/h, a new official record, albeit subsonic. It was just as well it was subsonic: computer modelling later revealed if *Thrust 2* had travelled just 11 km/h faster it would have taken off and – in all likelihood – smashed into 1000 pieces.

Things went quiet until the mid-1990s when several new LSR projects were announced, including one led by Rosco

'For my next trick, I'll set myself on fire.'

Craig Breedlove greets rescuers after smashing his *Spirit of America* jet car into a lake at something approaching 800 km/h (1964).

America's Craig Breedlove with the wingless fighter plane he argued was actually a car. Breedlove was the first to reach 400, 500 and 600 miles an hour on land. He held the LSR five times, as well as surviving the world's fastest car crash.

Gary Gabelich with the rocket powered *The Blue Flame*, the first car to achieve 1000 km/h. Gabelich was planning a supersonic LSR when he died in a traffic accident in 1984, crashing his motorcycle 'at great speed' into a truck.

McGlashan, an Australian drag racer. McGlashan hit 802.6 km/h for an Australian record but a supersonic run eluded him. Meanwhile Craig Breedlove had come out of retirement to have one last attempt at a supersonic LSR. This ultimately unsuccessful tilt involved Breedlove surviving the fastest accident in the history of motoring: basically an unscheduled right-hand turn at 1080 km/h.

The man who held the dubious honour of having the second-highest-speed accident also made a comeback: in 1966 Art Arfons had survived a smash at an estimated 982 km/h but even this didn't cure him of his addiction to speed. He had launched failed attempts at the LSR in 1989, 1990, 1991, 1994 and 1995. In 1996, at the age of 70 and with a triple heart bypass operation behind him, Arfons was trying it again in *Green Monster No. 27*. Unfortunately his low and lightweight challenger was beset with stability problems and vibrations and couldn't post a competitive speed.

Fourteen years after setting his LSR, Richard Noble was

also back. This time Noble was project leader rather than driver: his all-new, rear-steering, twin-jet monster was called *Thrust SSC* and had British fighter pilot Andy Green at the controls. *Thrust SSC* was a curious blend of rocket science and backyard engineering. During testing it left the ground at times, and engineers had to raise the rear suspension to push the nose further down. This caused huge vibrations and by the time they ironed out all the problems the car was literally falling apart with the strain of the speeds it was achieving.

Yet in the Black Rock Desert in Nevada on 15 October 1997, *Thrust SSC* produced an unmistakeable sonic boom twice in an hour to post the world's first official supersonic LSR. The

> **'For Britain, and for the hell of it.'**
>
> Richard Noble explains why he strapped himself to a jet engine to set the land speed record of 1020 km/h in *Thrust 2* (1983).

Richard Noble and his *Thrust 2*, plus a few earlier British LSR vehicles.

Noble's rear-steering *Thrust SSC* went supersonic. After achieving a world-record speed of 1227.952 km/h, British driver Andy Green said that rear-steering should never again be used on anything except forklift trucks.

official speed was 1227.952 km/h (763.035 mph) and *Thrust SSC* had covered a kilometre in less than 3 seconds. Neatly for the history books, Andy Green had scaled the Everest of automotive challenges almost exactly half a century after American pilot Chuck Yeager first broke the sound barrier in *Glamorous Glennis*, a Bell X-1 rocket plane.

There was an interesting contrast, however. When Yeager broke the sound barrier he was backed by the inexhaustible budget of the US Defense Department. *Thrust SSC* weighed ten tonnes and had the power of 1000 family sedans, yet it was controlled by a rear-wheel steering system that had been tested on a 30-year-old Mini in an abandoned car park.

Sledfoot

If there had been a practical (or military) reason to break the speed of sound on land it would have been achieved in the 1950s. Instead the quest has usually been undertaken by privateer adventurers, sometimes cobbling their machines together with second- and third-hand parts and mortgaging their houses to pay for the fuel. There was, however, one reason the military needed to go fast on land: to test jet, rocket and ejection systems. This led to a sort of parallel type of earth-bound flying, using military rocket sleds.

The first such sled was built by the Americans at Muroc/Edwards Air Force Base (California) shortly after World War II for V2 rocket research. Railway-style sleds are still in use at the Holloman Air Force Base, New Mexico, and speeds of 7000 feet per second (Mach 6) are believed to be attained on a regular basis, while speeds above 8900 feet per second have been occasionally achieved. That's more than 6000 mph, or 10,000 km/h. Acceleration forces are said to be high enough to turn a human passenger to jelly. Nonetheless, chimpanzees and even humans have been used as test subjects in lower-speed runs.

Dr John Paul (Col.) Stapp, a surgeon and former US Air Force officer, earned the title 'The Fastest Man Alive' when he rode the *Sonic Wind I* rocket-propelled sled in December 1954. Dr Stapp accelerated to 632 mph (approximately 1018 km/h) in 5 seconds, an unprecedented speed on land for that era, though not exactly achieved within the land speed record rules. Stapp then survived the greatest controlled human deceleration recorded up to that time, when he was hauled to a stop in just 1.25 seconds. The wind blast was equivalent to an ejection at supersonic speed at high altitude, while the braking force was more than 40G, with Stapp's bodyweight momentarily exceeding 3 tonnes. Dr Stapp completed 27 sled tests, suffering a string of retinal haemorrhages, plus cracked ribs and two broken wrists. His efforts helped improve aircraft seatbelt, helmet and ejector seat technology. In his later years Stapp also became a respected safety advisor to the automotive industry.

> **'I'm an artist, the track is my canvas, and the car is my brush.'**
>
> Line credited to British racing driver Graham Hill (1929-1975). Hill didn't pass his driving test until the age of 24, but became world Formula One champion twice – in 1962 and 1968. In 1996 his son Damon became the only second-generation Formula One champion.

Le Mans, 1927, and a couple of Bentleys get away early. It was to be the first of four consecutive victories for the marque, followed by bankruptcy.

> **'I don't smoke, and neither do my cars.'**
>
> Enzo Ferrari's legendary explanation as to why he would not accept cigarette advertising on his racing cars when it was first proffered in the late 1960s. It was true that *he* didn't smoke.

The 1930s Grand Prix madness was fuelled by the Nazis, who wanted to show German superiority. The result was cars such as the 1937 Mercedes-Benz W125, which put a positively frightening 482 kW (646 bhp) through relatively narrow cross-ply tyres. The W125's power output wouldn't be exceeded by Formula One cars until the early 1980s.

> Ford US announced the revival of its legendary Le Mans-winning GT40 coupe in 2002, then discovered it hadn't registered the name and couldn't use it. The car became simply the Ford GT.

Top Formula

When you drive a modern Formula One car through a chicane, it's like having a full suitcase dropped on one side of your head, then the other. Hitting the brakes is, to quote driver David Coulthard, the equivalent of having the back of your helmet whacked with a sledgehammer.

Pulse rates can hit 200 beats per minute, the very limit of the heart's capabilities, and in one race Finnish driver Mika Hakkinen sweated away more than 4 kg.

This is the modern, 'safe' world of Formula One, where race-drivers are far less likely to die but are giving their bodies such a fearsome workout they are all on the way to degenerative arthritis. What makes Formula One so hard on the body is that the races are long (up to two hours), the cockpit is fearsomely hot, the forces of cornering and braking are harsh enough to make an untrained driver black out, and the concentration required is unremitting because the consequences of a lapse at 350 km/h can be fatal.

> 'That day I did things that I had never tried before at the wheel and I wouldn't ever want to do again.'
>
> Juan Manuel Fangio of the German Grand Prix at the Nürburgring in 1957. An unscheduled pit-stop had left Fangio's Maserati well behind the Ferraris of Brits Mike Hawthorn and Peter Collins; in his pursuit on the 174-turn, 22.7 km track Fangio broke the lap record no less than 10 times. The Master eventually caught and passed the Ferraris and won the race, all the time holding himself in place with his knees because his seat had broken. Don't try it at home.

Michael Schumacher, seven times World Formula One Drivers Champion, earned a reputed US$60 million in 2004. As a two-time world champ in the early 1960s, Australia's Jack Brabham operated a service station in England to make ends meet. Further proof of the contrast: this shot of Brabham enlisting wife Betty to clean his goggles during a mid-race pit stop. If the photo wasn't so tightly cropped we'd probably see Jack's mother changing the front right wheel.

Clark of the course: Scotland's first world champion Formula One racer, Jim Clark, is still revered by many as the greatest race driver of them all. He was the miracle man, the one who pulled his car back from impossible spins and slides, who won almost everything he entered, who never did himself the slightest injury in all his years of motor racing. That was until April 1968 when something went tragically wrong in some minor Formula Two race that a man of Clark's talents should not have even bothered with.

Formula One was established as the premier international open-wheeler formula in 1950 and the first true world championship came into being. Although the sport was more organised than before the war, it wasn't a whole lot more interested in driver longevity. During the first decade of Formula One, 21 drivers were killed.

In the 1960s the figure was even worse, with 27 deaths, many of them caused when cars ran into unfenced trees, poorly designed fuel tanks exploded on impact, or injured drivers were left stranded a long way from medical support. By the 1970s, however, the death toll fell to a neat dozen, partly due to the efforts of Scotsman Jackie Stewart.

The idea that his favourite sport should be safer came to Stewart one day in 1966 while he was trapped in a crashed BRM. Although he was soaked with leaking fuel and any

Late victory: Paul Warwick not only won the 1991 British Formula 3000 Championship posthumously, he won his final race in the same unfortunate fashion. When Warwick's buckled car blocked the track and forced the Oulton Park Gold Cup race to be stopped, the results were taken back to the previous lap, which Warwick had led. The victory gave him an unbeatable points lead in the championship. It was a component failure that caused the fatal accident.

errant spark was going to end his life, the marshals had no tools to free him. From then on Stewart campaigned for safer tracks, safer cars, better-equipped marshals and proper medical facilities at every circuit. He argued that he and his colleagues were being paid for their driving talent, not for needlessly throwing away their lives. It may seem like common sense, but many dubbed Stewart a coward and said, in effect, that safety had no place in motor racing.

Stewart was able to show he wasn't a coward by beating most of his contemporaries in almost any conditions, and gaining three world championships along the way. His high-pitched, unceasing voice was heard loudly and clearly as a result. During the 1980s the death toll dropped further (to seven) and between 1990 and the end of 2005 the figure would have been zero except for one appalling weekend. Austrian Roland Ratzenberger and Brazilian Ayrton Senna died in separate collisions at the 1994 San Marino Grand Prix.

The reduction in fatalities has been due to constantly improving safety equipment, including on-board fire extinguishers, carbon-fibre 'tubs', capable of absorbing huge impacts without crushing, better harnesses and helmets, and smarter circuit design. They don't put trees on the apexes of corners any more, for example. Oddly, though, as cars have become safer they have also become much harder on the drivers due to the improved aerodynamics, more powerful

Perhaps the best demonstration of Jim Clark's uncanny abilities: in the 1965 Formula One season there were ten races. Clark turned up for nine and won six. He had a good excuse for the GP he missed: he was in the US becoming the first foreign driver in 49 years to win the Indy 500. Of the three GPs he didn't win: he set pole, fastest lap and led in Italy, led at Watkins Glen (USA) and was on pole for the final race at Mexico, but in each case circumstances contrived to keep him from winning.

Australia's Alan Jones was for a brief period the toughest, fastest racing driver in the world. He was rewarded with the 1980 World Formula One Championship. Despite his no-nonsense demeanour, Jones admitted he always wore red underwear while driving because it brought him good luck.

> **'The danger sensation is exciting. The challenge is to find new dangers.'**
>
> Ayrton Senna, talking in 1989. Unfortunately the brilliant Brazilian found the ultimate danger during the San Marino Grand Prix in Italy in 1994. He lost control while leading and fatally collided with a concrete wall.

engines, stickier tyres and tall-sided bodywork that improves protection but cuts air-flow to the driver down to, roughly speaking, zero. Even when the weather is cool, the drivers lose 2 kg in sweat. Hakkinen losing more than 4 kg from his 72.5-kg frame (during the 1999 Malaysian Grand Prix) was the result of unusually high temperatures and the fact his drink bottle, which would normally allow him to replace some fluids, was broken.

Long-time Grand Prix doctor Professor Sid Watkins says a pulse rate of 200 means the heart 'is beating that quickly it doesn't have enough time to fill properly so you are getting to the limit of your cardiac reserve'.

New Zealand's Chris Amon spent 800 km leading Formula One Grands Prix but never crossed the line first. Britain's Peter Gethin led for just 11 km and won the 1971 Italian Grand Prix. He crossed the line just one hundredth of a second ahead of Ronnie Peterson, making for the closest Grand Prix finish of all.

> The wheels and tyres of a Formula One car revolve at approximately 2800 rpm down a long straight, creating a strong gyroscopic effect.

'And these cars vibrate because they have very little suspension, if any, and so they get a lot of repetitive trauma to the spine and neck, which results eventually in some degree of degenerative arthritis.'

By most estimates, a normal, fit road driver would last between two and five laps in a Formula One car before his neck hurt so much he couldn't hold his head up.

During qualifying for the 1997 Imola Grand Prix, German driver Heinz-Harald Frentzen recorded a force of 5.99G under braking. Momentarily, his body weighed six times its normal amount and the blood supply to his retinae was restricted, distorting his vision. Yet the cars are now so safe that Brazilian driver Felipe Massa crashed in Montreal in 2004 with such terrifying force that the in-car instruments registered 200G. He hurt his elbow.

> 'Losing one ear . . . never really bothered me all that much. Just the opposite – now it's easier to use the telephone. The receiver is a little bit closer to the eardrum, which comes in handy when you are making a long-distance call.'
>
> Niki Lauda reveals his practical streak when talking about his motor racing injuries in the book *Second Time Around* (1982).

Always fast, sometimes furious, and very, very expensive: modern Formula One. An analysis by *F1* magazine suggested Toyota spent US$499.05 million racing two cars in the category in 2005. Here Spain's Fernando Alonso leads in a Renault in 2006.

CHAPTER ELEVEN

It seemed like a good idea

They were the innovations supposed to revolutionise the car world but which went the way of the eight-track cartridge. Ideas such as four-wheel steering, pop-up headlights, digital speedos and, well, the eight-track cartridge. Join us for a walk down this boulevard of broken automotive dreams. But mind your step. It may be a wrong one.

Badge engineering

The idea works in fridges and stoves: you take a generic product, add some unique handles and a new chrome moulding or two, and flog it under your own brand name. But cars are not whitegoods and customers usually see through the ruse. Despite this, some remarkably stupid examples have been foisted on the public in recent decades. They included Nissan Australia's Bicentennial gift to the nation – a bog-standard 1988 Falcon ute with Nissan badges – and, a year later, the Toyota Lexcen. This was a rebadged and almost

> 'While the Skycar looks really cool, it doesn't do much sky-ing, nor is it capable of much car-ing.'
>
> *Fortune* magazine (December 2004) comments on what Dr Paul Moller calls the greatest advance ever in personal transportation, the Moller Skycar.

IT SEEMED LIKE A GOOD IDEA

An early petrol-electric hybrid, from the Briggs & Stratton Corporation. The extra rear wheels were needed to support the weight, though it was claimed to be practical. Mind you, it was claimed to be elegant and stylish too.

unchanged Holden Commodore. The name Lexcen (after yacht designer Ben Lexcen) seemed specially chosen to cause confusion with Toyota's Lexus model, which had been unveiled overseas earlier in 1989. Furthermore, Toyota had successfully built its international reputation on quality and reliability, and they were two things the late 1980s Commodore had in short supply.

Digital speedos

People paid huge premiums for these in the 1980s and quickly discovered they didn't like them – and had to replace the whole car just to change them. The main problem: it takes more concentration to read an ever-changing number than it does to check that a conventional needle is hovering within a certain range. Added to which, the first digital dashboards were famously unreliable. Good riddance, said most.

> 'All things considered, the DeLorean has a lot in its favour. The appearance is striking, it's very sporty and rewarding to drive, yet comfortable and relaxing on long journeys. And on top of this it should last forever and a fortnight'.
>
> *Motor Trend* magazine gets it wrong by roughly one 'forever' when discussing the disastrous Irish-built US sports car (May 1981).

In 1994 Saudi Arabian traffic police reported arresting more than 500 'stunt drivers' and 15 people who had been 'teasing road users'. In that same year 186,640 vehicles were towed away in Paris – 180,000 of them because they were illegally parked.

Two concept cars built around Australian Ralph Sarich's Orbital two-stroke engine: Ford's sub-B and GM's Ultralite from the early 1990s. The US majors seriously considered building the engine, then seriously unconsidered it when the small and powerful three-cylinder failed to meet durability targets.

Pop-up headlights

No need for the bug eyes of a Bug-Eyed Sprite when you've got headlights that can crouch and hide in the bodywork to provide a clean aerodynamic profile during the day, and can pop up to meet the minimum height legislation for headlights at night. So popular were these little hooded eyes that at one point some luxury sedans had them, even an everyday hatch or two. They increased cost and weight and didn't do much else. And, when new headlight technology allowed 'lay down' or ultra-low-profile headlights (and exotically shaped ones at that), the slight arguments in favour of pop-ups became slighter still.

The electric car

Not since the heyday of the electric cars in the early years of the twentieth century, has anyone turned a buck making battery-powered road vehicles, despite many attempts. Batteries as we now know them simply aren't an efficient power source for cars, trucks or buses. Furthermore, they are heavy, so dragging them around increases your energy consumption, and makes for horrible possibilities in a crash. Yet the electric car still has a magical aura in the eyes of certain people, who see it as the solution to all environmental problems. Alas, it isn't.

> 'When a person is driving an automobile there is a sense of complete control. It responds without a moment's hesitation to whatever we tell it to do, even to its own demise, even to our own demise. There's no other situation in our lives where that happens – and we love it.'
>
> Dr Robert Guenther, Detroit Medical Center, interviewed in the television documentary *Crash Science* (1998).

Although some car names to come out of South Korea sound like skin diseases (tell me, doctor, it's not Nubira is it?), models have also been named after popular pub games, such as the Daewoo d'ARTS. Other Korean automotive monikers have included: Mapsy, Royal Duke, Tico, Spagon (Daewoo); Credos, Brisa, Potentia, Avella (Kia); Marcia, Presto, Tiburon, Dynasty, Galloper (Hyundai) and New Family (Ssangyong).

STEP ON IT!

General Motors spent $1 billion developing the Impact (or EV1) electric car in the 1990s, a figure partly subsidised by the Clinton government. Built mainly for California (where laws required a certain percentage of 'zero emission' vehicles), it attracted waiting lists but GM didn't increase production, would only lease rather than sell the vehicle and crushed many at the end of the lease. There were conspiracy theories suggesting GM had set out to fail.

Steam Power

The battle between steam engines and petrol engines was fought and decided generations ago, yet every few years a new 'revolutionary' automotive steam engine turns up. One of the more famous attempts to re-fight the battle was in 1968, when William P. Lear employed 150 engineers to test

The future is here: All 1978 Chryslers were available with a 40-channel CB transceiver. Do you copy that? Probably not.

190

every known steam design, then develop an entirely new variation with 'space age' technology. Lear was the man who brought us the Lear Jet and, yes, the eight-track cartridge. His eventual steam engine, a lightweight turbine (with a claimed 186 kW), was tested in several cars but no major car company took it up. Melbourne engineer Ted Pritchard and Sydney architect Gene Van Grecken also developed 'revolutionary' steam engines around this time with a similar lack of commercial success.

> 'I first drove a Volvo in America when I was over there on business. I haven't looked back since.'
>
> Australian fashion designer Lisa Ho quoted in a Volvo press release (2001). It obviously didn't take her long to acquire the customary driving style.

Four-wheel steering

Honda stunned the car world in 1987 with a Prelude coupe that steered at both ends, something never before seen in a volume-produced car. *Wheels* magazine named it 'Car of the Year' and called 4WS an advance comparable to the disc brake and automatic transmission. Mazda followed with an even more sophisticated version of the same concept and predicted that by 1995 half of all new cars would be fitted with 4WS. The Europeans argued 4WS was a piece of frippery from cocky Japanese engineers rather than a real advance in cars. Yet soon after we saw many European concept cars with all four wheels akimbo. The Euros never got serious about putting their systems on production cars, however, and the Japanese soon gave up. They realised that engineers could achieve many of the same gains in ways that were more cost-effective, lighter and created less work for wheel alignment specialists.

Four-wheel steering has been tried many times over the years. This is an English attempt from the 1960s that allowed a car to turn in almost its own length.

Honda's production four-wheel steering system caused huge excitement. For a while.

Toggle style blinkers

There are some things you just can't change and the blinker switch is one of them. Like the speedo dial, any attempts to update the good old-fashioned original has proved unwelcome. Jaguar tried and abandoned a 'clicker' blinker arrangement and Japan brought us no end of binnacle-mounted atrocities in the 1980s. But, to a man and woman, we said: 'no thanks, give us what we had'.

Turbine engines and more

It didn't quite create the excitement of the Wankel, but around 1950 there was a strong belief among some that we were all going turbine. Rover built the first turbine-powered car; then Chrysler produced its turbine 'car of the future' and sent it around the world. Neither went into production.

Four wheels good . . . eight wheels better. This 1911 Octocar was claimed to cut down on delays caused by punctures, though the maker, Milt Reeves of Indiana (USA), also argued that the extra wheels made the vehicle safer and more comfortable. And so few obvious disadvantages, too.

IT SEEMED LIKE A GOOD IDEA

Canada's Dr Paul Moller has spent almost his entire life trying to build a flying car. This is an early attempt (from 1971), using four engines to spin the outer ring. 'What I've tried to do is . . . simplify the machine to the point where it can be mass produced at the price of an automobile and build in the safety factor necessary so that a housewife can fly to the local market,' he said. Thirty-five years later, Dr Moller was still trying to sell his much improved, though still not housewife-to-local-market-ready, Skycar.

Almost every other possible power source, from engines running on mulched grass to nuclear power, has been exhibited by major car makers over the years. And there have been hundreds, if not thousands, of different petrol engines patented since the early days of motoring.

Pin code door locks

Who needs a conventional key when you can have a door that looks like a pocket calculator and requires you to simply thump a few numbers to get in? Nissan thought it was a great idea during the 1990s and applied it to the Maxima

The early 1970s brought a range of experimental micro cars from the big manufacturers, including GM, which built this '512' two-seater, steered with a tiller. The company claimed that although it would not put any of its micro-cars into production, it had learned important lessons for future use. It then continued building enormous fuel-guzzling tanks.

Another GM experimental car, this one from the 1980s and designed to lean into corners. The claim was that it could travel 'up to 200 miles on a gallon of gas'.

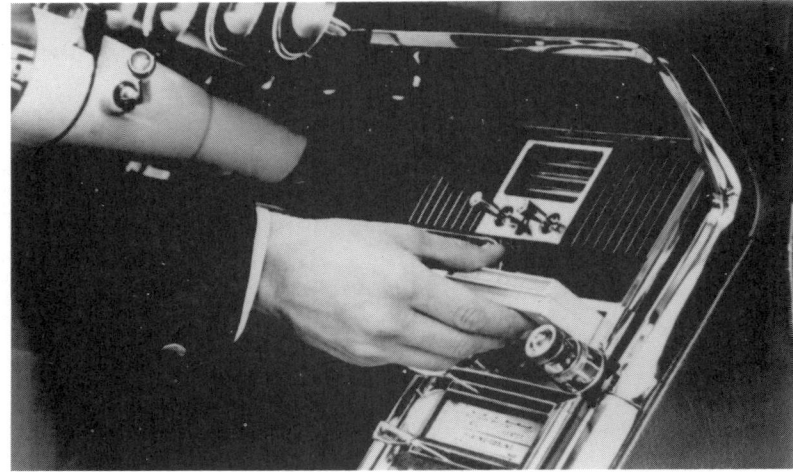

The eight-track cartridge, as fitted to the 1966 Ford Thunderbird. It had lousy sound quality, was unreliable and couldn't be fast-forwarded or rewound – but it was the first practical way of playing recorded music in a moving vehicle. The vastly superior Compact Cassette player (designed by Philips) was first seen in Pontiacs in 1971, though the eight-track remained common until the early 1980s.

Motorised roller skates, as demonstrated by Alphonse Constantini in 1906 and supposedly capable of 70 km/h. Scary stuff, particularly when one or both stalled at speed.

(mind you, Nissan thought the 1990s Maxima was a great idea too). The concept quickly established an unpopularity it has successfully maintained until this day. Yes, there now are some keyless entry systems, but none with a keypad on the door. Nissan was allowed to keep that one all to itself.

Radical safety systems

There have been many strange ideas about how to make a car safer, including giving cars needle-sharp noses so that a head-on collision was impossible (shame about a side-on). Two-wheeled road-users are particularly vulnerable, and Troy Jackson and Joseph Leak of Philadelphia certainly had this in mind when they registered their safety system in 1997 (patent 5,593,111). As insanely complicated as it is gloriously daft, it involves the rider wearing a large folded-up wing or, alternatively, a parachute. A sensor – which you'd want to be pretty reliable – detects an imminent accident and blasts the rider skyward, at which point the wing or parachute is 'rapidly deployed,

IT SEEMED LIKE A GOOD IDEA

> The JiangLing Landwind, displayed at the 2005 Frankfurt Motor Show and slated to be the first Chinese vehicle on sale in Europe, underwent crash tests and recorded a score of zero. German automobile club ADAC (a Euro NCAP-approved tester) announced 'In our 20-year history no car has performed as badly.' Presumably the Landwind is aimed at BASE jumpers and ultra-light aircraft devotees.

thereby causing the velocity of the ejected rider to be reduced and providing the further benefit of lifting the rider away from the crash site'. Our advice: don't have an accident in a tunnel. Other patented cycle safety systems involve suits that inflate to protect in the event of a collision or, in one case, a box-like network of airbags that instantly erects around the bike to form what looks like one of those kiddie jumping castles.

> **'The Ignited Colours of Benetton.'**
>
> London's *The Sun* chortles in headline type when a fuel leak transforms the Benetton Formula One pits into an inferno (1994). Those wacky Brits at 'The Current Bun' found similar merriment in 1996 when the Ligier of Brazilian driver Pedro Diniz turned into a fireball: 'Diniz in the Oven'.

Patently silly: the motorcycle wing. This is the working drawing accompanying patent 5,593,111.

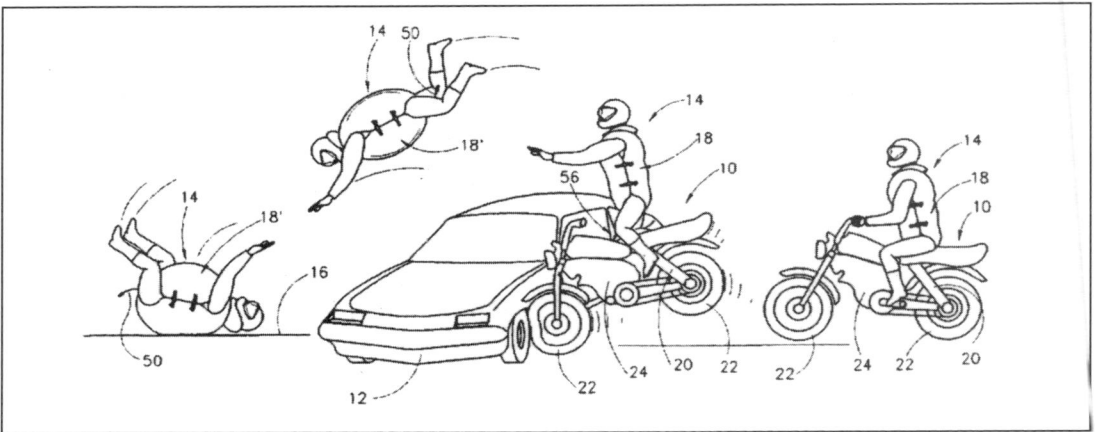

Do-it-yourself Michelin Man: patent drawings for a wearable airbag.

No steering column intrusion here. An early Australian experimental safety car, based around a Falcon, had the novel solution of putting a steering wheel on each side of the driver's seat.

> Renault recalled 45,000 Meganes, including 500 Australian vehicles, in 2004 because the brakes could be inadvertently applied from the passenger side. The right-hand drive conversion of the model retained some of the brake pedal mechanism under the left footwell, and sufficient pressure on the floor – by a nervous passenger, for example – could activate the brakes.

Look mum, no side mirrors. Ford was sure the rear-view periscope was going to replace the rear view mirror by the year 2000. Interesting dead-ends that have actually made it into production include the Procon Ten cable safety system (Audi), double glazed windows (Benz S-class), solar panel sunroof (Mazda 929), and such aero devices as bonnet 'air splitters'. And did we mention inflatable roof racks?

Rotary hysteria

The search for the miracle car engine has at times rivalled the earlier quest for the philosopher's stone. Thousands of alternative engines – most of them rotary designs – have been developed and patented by a determined breed of engineering alchemists. But none has caused more excitement than the rotary engine developed by Dr Felix Wankel of Germany.

In the late 1950s, after 30 years' work, Wankel launched his engine concept and almost from day one experts were saying that it would make the conventional piston engine obsolete. The Wankel promised light weight, smoothness, simplicity, cheaper production and the ability to push out a huge power output from a small displacement.

The first Wankel-powered car was the NSU-Spider. Some complained that the central driving position was a little too high, yet the head could still be kept warm with a well-designed hat.

The prototype weighed only 11 kg but developed 22 kilowatts at 17,000 rpm. It consisted of an inner rotor (triangular in shape) that followed an unusual path within a stationary outer housing. Car makers were mesmerised by the possibilities, and by the 1960s the public was whipped into such a frenzy that engine technology suddenly became front-page news. General Motors, Rolls-Royce, Mercedes-Benz, Citroën, Mazda and even Russian car maker Volga were among those who signed up for a licence to produce Wankel rotary engines for fear of being left behind forever.

The Wankel engine made its production debut in the 1964 NSU-Spider Sports, then appeared in 1967 in the radically designed NSU Ro80 sedan. Wankel engines for boats, aircraft, motorcycles, generators, trucks, lawn-mowers and hovercraft were designed and in many cases built. Yet all the companies working on the Wankel engine were faced with severe durability problems, heavy fuel consumption and excessive exhaust emissions.

Citroën produced a few production cars before saying enough is enough, but most others abandoned their Wankel programs before they got that far. GM is said to have spent over US$1 billion on the whole affair without selling a single rotary powered car. Mazda, however, continued with such determination that it literally sent itself bankrupt. NSU was forced into an amalgamation with Audi by the cost of

Inside the Wankel engine: as it moved around the housing, the rotor created sealed chambers, which progressively varied in size. This allowed the induction, compression, power and exhaust 'strokes' to be completed without the inertia losses associated with having a piston constantly change its direction of travel. It was simple, elegant and capable of bankrupting anyone who tried to produce it.

warranty claims on the Ro80. By the 1970s, it was only Mazda (newly bailed out by Japanese banks and Ford) that was pushing forward with the sale of rotary vehicles, finding ever more problems along the way and never winning the masses over to the concept.

Felix Wankel died in 1988 at the age of 86 years. By that time Mazda had built 2 million vehicles powered by Wankel engines – a decent tally, but nothing compared with earlier predictions.

Citroën was determined not to build a rotary-powered car unless it could build a body ungainly enough to compensate for the Wankel's elegant simplicity. Just 267 examples of the M35 were built; Citroën bought most of them back and destroyed them.

> The late Felix Wankel, who gave his unlikely name to the only successfully mass-produced rotary car engine, never learned to drive. Nor did Dr Luigi Fusi, who designed Alfa Romeo's highly successful race cars of the 1930s.

STEP ON IT!

> **Crap names:** the follow up to Ford's EB Falcon was to be the EC until the eleventh hour when it was pointed out that, in some parts of Queensland, EC was the expression for earth closet or toilet. It became the Falcon ED. When Toyota launched the MR2 in France, the local pronunciation (em-er-der) sounded too close to *merde*.

Legendary stuff

Car makers and oil companies have kept all sorts of radical and useful technology from us over the years, in keeping with their undying commitment to evil. For confirmation, just ask any conspiracy theorist, or Uncle Bert, who will no doubt tell you how a friend of his father invented a special device that halved fuel consumption. You know the rest: a man from General Motors, Mobil or the government turned up to buy the technology and took away all the prototypes and drawings and was never seen again.

The famously unsuccessful Ford Edsel was promoted in 1957 with the line 'They'll know you've arrived when you drive up in an Edsel'. Unfortunately, you'd know you'd arrived too, because you'd hear the people sniggering.

IT SEEMED LIKE A GOOD IDEA

Mazda announced that its 1999 Neospace concept car was designed with the twin principles of 'Contrast in Harmony' and 'Dynamic Strong Wheel Oriented Design'. If nothing else, the company's press release writers had shown they could tread the fine line between rubbish and utter rubbish.

The miracle engine/carburettor/fuel-saving device referred to by Uncle Bert and many others has a long history in popular myth. One version has a long-serving GM or Ford employee receiving a car as a retirement gift and discovering that the fuel gauge scarcely moves. Another has a man (or woman) buying a car in another city and driving it home without ever having to fill up. The next part of the story usually involves the car being stolen back by the company (which has, of course, accidentally released one of its fuel-

The controversial Chevrolet Corvair. In 1960 the reigning Formula One world champion, Jack Brabham, was pressing hard on a wet road, trying to make a race meeting in time. When his Corvair hit a patch of wet leaves, he had a size thirteen spin and the story grew to legendary proportions. If the world champion couldn't control the Corvair, what hope was there for the average driver?

201

> Lada cars are manufactured in Togliatti, a Russian city named after an Italian socialist.

saving experimental cars never intended for the market), or has the owner waking up in the middle of the night and looking out the bedroom window to see mysterious men tinkering with the engine. Nothing appears wrong the next day – except the fuel economy is now back to normal. Another variation has a man in a suit turning up and making a fabulous offer for the car.

The reality is that if a car company had any such device up its sleeve, there's at least a fair chance it would express its innate evilness not by suppressing it but by exploiting the advantage. The savings in running costs could be shifted back into the price of the car. Governments, likewise, could save billions of foreign exchange from any miracle fuel saver and apply the fuel taxes somewhere else. Even if a fuel company had secured the invention, it may well decide that supplying the technology to the tens of millions of cars built

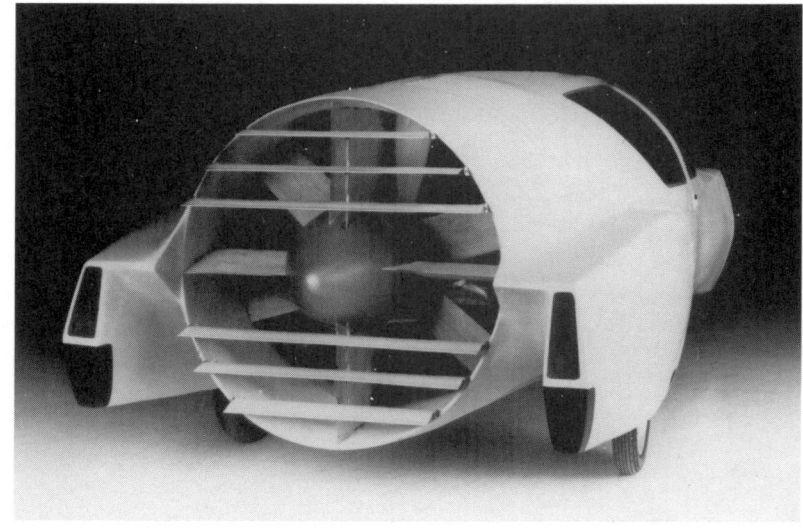

The Bede Fan Car, from 1980. Another brilliant idea claimed to improve fuel economy and deliver a whole lot of other advantages, but somehow never quite making it to the shopping centre car park.

IT SEEMED LIKE A GOOD IDEA

Not as silly as it looks: the RoboRoo is a kangaroo test dummy used by Holden to 'study a uniquely Australian road safety issue'. Well, actually, it is as silly as it looks.

each year was at least as profitable as selling fuel, the price of which it could crank up anyway to compensate for lower sales. But hey, where would that leave Uncle Bert?

Gunning for victory

In 1998 Reuters relayed a report from the newspaper *Iltalehti* stating that a 42-year-old Finnish man had been charged with manslaughter after strangling his 68-year-old mother for turning off the television during the Luxembourg Formula One Grand Prix. Whether the man was able to see the rest of the race was not mentioned. However, surveys showed that one million Finns, or one quarter of the country's entire population, did watch their compatriot Mika Hakkinen take the chequered flag that year. And perhaps the extent of the national Formula One fervour was shown by matricide during a motor race counting only as manslaughter.

More bizarrely, there is at least one example of a gun being pulled by a competitor during a motor race. The incident involved Wendell Scott, a driver from the US state of Virginia who, although black, was determined to compete

> 'People who come up with "It may not work" or "What are we going to do if it fails?" do not have the credentials to be businessmen. If there is only a 1 per cent chance of success, a true businessperson sees that 1 per cent as the spark to light a fire.'

Former boss of South Korea's Daewoo company, Mr Kim Woo-Choong. By his own definition Chairman Kim was a true businessman. In just over 30 years, he turned a US$5000 loan into a US$51.8 billion (62 trillion won) loan, along the way producing dull but cheap cars and writing the book *The World is Big and There Are a Lot of Things to Do*. When Daewoo collapsed in 1998 he went on the run. He was finally arrested in 2005.

203

> 'I firmly believe that a man can do whatever he sets his mind to doing . . . Frankly, I'm so rabid on this idea that I can say to you with a straight face that I believe that if a man believed hard enough, and excluded everything else from his life, he could walk on water. I mean it. I'm not kidding.'
>
> Sir Stirling Moss quoted in the *Encyclopaedia of Auto Racing Greats*, by Robert Cutter and Bob Fendell (Prentice-Hall, 1999). Obviously Sir Stirling never really wanted to win the Formula One World Championship.

in the notoriously redneck NASCAR category. Scott was variously banned or otherwise prevented from competing during the 1950s, but tenaciously stayed on the case. In Jacksonville, Florida, in 1963, he became the first and only black man to win a major NASCAR race.

Problem was that officials feared a riot if Scott walked into the winner's circle, let alone exercised the victor's right to kiss the promotional girls, who were white. The solution was to hand the trophy to another driver, Buck Baker, who appeared to finish two laps behind. A few hours later officials announced there had been a scoring error and awarded the victory to Scott, but by then all the grandstands were empty.

Scott, however, did not take his harsh treatment lying down. His son Franklin told an interviewer that in 1962 a driver named Jack Smith was so enraged by Scott's qualifying success he threatened to knock him off the track.

'On the pace lap he pulled up beside Daddy and started pointing his finger at him. We didn't know it but Daddy had his gun with him and he pulled it out and pointed the gun back. We never had trouble with Jack again.'

NASCAR inducted Wendell Scott into its Hall of Fame in October 1998. The sense of timing was no less insulting than it had been when he had been awarded his first victory. By 1998 Scott had been dead for 29 years.

In 1999 Toyota Australia chose to name its new family sedan after the Centaur, the mythical Thessalian half horse-half man. Then it was pointed out *Centaur* was also the name of an Australian hospital ship sunk by a Japanese submarine in 1943 despite being illuminated with prominent red crosses. The death toll was 268, mostly nurses on their way to Port Moresby to pick up sick and wounded troops. Toyota Australia quickly became a name-dropper. The car became the Avalon.

CHAPTER TWELVE

Simply the best

'Whatever is rightly done, however humble, is noble,' said Henry Royce, in between building some of the largest, most expensive and distinctly unhumble road machines in the history of this particular planet.

Royce, the engineering half of the Rolls-Royce duo, failed to proceed in 1933, but by that time his cars had become a legend and the brand a byword for quality. Such a byword, indeed, that the brand managed to keep its sheen during the 1980s and 1990s, by which time it had been long outclassed in every measurable way by the offerings of the big German companies.

Frederick Henry Royce co-founded the Rolls-Royce company in 1906 with the aristocratic Charles Stewart Rolls, who was to handle the sales side. Rolls was not an engineer but was captivated by anything mechanical. He was one of Britain's first motor car dealers and competed in many early motor racing events as well as being a pioneer aviator. But

> 'Ford and the world Fords with you, Rolls and you roll alone.'
>
> Line attributed to F. E. C. Smith, aka Frederick Edwin Smith, 1st Earl of Birkenhead, Viscount Furneaux of Charlton, Viscount Birkenhead of Birkenhead, Baron Birkenhead of Birkenhead (1872–1930). Needless to say, he owned at least one Rolls-Royce: a 1914 Silver Ghost with Tourer body by Cockshoot.

This 1905 Rolls-Royce, the Legalimit, had a V8 engine but the top speed was restricted to 20 mph (about 32 km/h). The fact that it was smooth and quiet wasn't quite enough: nobody was going to buy a V8 to travel at that speed. Only three were sold and Henry Royce eventually bought them back.

No such restrictions: the fast, powerful and fearfully expensive 1925 Rolls-Royce Phantom 1 roadster, with separate windscreen for the rear seat passengers, was the perfect embodiment of the Roaring Twenties.

in 1910 the tail fell off his Wright Flyer, affording Rolls the unfortunate honour of being the first Englishman to die in an aircraft accident.

Some may assume that the Rolls-Royce has always been held as the 'best car in the world', but through the years there have been many claimants. Before World War I a strong contender was fellow British marque Napier, but Rolls-Royce's Silver Ghost soon proved to be smoother, quieter and more graceful. It became a favourite of royalty, with Tsar

> **Flamboyant American millionaire Tommy Manville Jr, heir to an asbestos fortune, worked his way through 13 marriages and 11 Rolls-Royces.**

Bentley, top hat, light tie, and tails. This man seems unaware the brand would be owned by Volkswagen before the century was out.

Nicolas II setting the pace, buying two in 1913. During World War I the Ghost was used for ambulance work and proved just about the most reliable machine of any type used by the British or French armies.

Add the visual impact of that huge Parthenon-in-silver that was its radiator grille, and the Rolls-Royce was soon the

The large and brilliantly engineered Duesenberg was the preferred conveyance of many obscenely wealthy Americans. This one was photographed in 1935, but the company was already in trouble. The last 'Doosey' would be built in 1937.

Bugatti's Royale, arguably the most expensive car of all time, and certainly one of the largest. This is a 1931 example. Just six were made.

> **'It's not the company's policy to comment on owners' tastes.'**
>
> The reported response of a Rolls-Royce spokesman when questioned on the wisdom of Beatle John Lennon having his Crewe creation painted in psychedelic colours (1967).

benchmark against which others from around the world were judged.

American efforts to secure the lucrative mantle of 'most luxurious machine on the road' were numerous. Those that staked the strongest claim included the Stutz Motor Car Company of Indianapolis, which followed its famous Bearcat speedster of the World War I era with fast eight-cylinder tourers in the 1920s and a series of superb limousines in the 1930s. The Great Depression, however, was rather cruel to companies making cars that cost more than the average worker earned in a lifetime. The crash ended the Stutz line, as it did that of several other top-line Americans. Also on the automotive breadline were Duesenberg (so opulent its name survives in the expression 'it's a doosey') and Pierce-Arrow, whose eight- and 12-cylinder models had been considered by many the most prestigious on the market. Pierce-Arrow's 6-66 Raceabout, offered between 1910 and 1918, boasted an engine capacity of 13.5 litres. A Model T made do with 2.9 litres.

The original Packard company limped through into the 1940s but never regained its pre-World-War-II reputation as a maker of some of the world's finest cars, while Cadillac dropped its V16 model at the end of the 1930s and, post-war, concentrated more on gadgets and gizmos than ultimate refinement.

The French brand Bugatti launched its 12.7-litre Royale model to out-Rolls-Royce Rolls-Royce, but only six were built. It might have had something to do with the price: three times as high as the dearest Roller. Any potential competition from Bentley was snaffled in 1931 when the highly regarded but perpetually cash-strapped British company was bought by Rolls-Royce itself. Other Europeans made a solid tilt for the title: the French-Spanish marque known as Hispano–Suiza or, colloquially, Banana Squeezer (it folded in 1938) and Belgium's Minerva brand, which faded away around the same time. France's Voisin, Delage and Delahaye marques also shone for a while.

Before their merger in 1926, Mercedes and Benz both built

Maybach, the pre-WWII German brand, returned in the twenty-first century. This time around it was built by DaimlerChrysler and was essentially an uber-Mercedes aimed squarely at the new Rolls-Royce Phantom. Reviewers used the words 'big' and 'imposing' a lot more than they used the word 'beautiful'.

The first new Bentley produced under VW control was the Continental GT coupe. It was an instant hit, and became a favourite of bling-laden rap artists.

'I think that cars today are almost the exact equivalents of the great Gothic cathedrals. I mean the supreme creation of an era, conceived with passion by unknown artists and consumed in image, if not in usage, by a whole population which appropriates them as a purely magical object.'

Roland Barthes (1955) in his famous dissertation in praise of the Citroën DS. Sadly, Barthes died in Paris in 1980 from injuries suffered when struck by a motor vehicle.

cars favoured by royalty and millionaires, as did Germany's Horch company (it was absorbed into Auto-Union, now Audi, and did not survive World War II in its own right). In 1964 Mercedes-Benz made its most audacious play for the 'best car' tag with the monstrous Grosser 600 Pullman, available with six doors. It proved a mechanic's nightmare, though very popular with South American dictators. The Pullman theme was revived in the twenty-first century with the long wheelbase version of the Maybach, built by DaimlerChrysler and named after one of the most expensive German cars of the 1930s. This time the longest version was just under 6.2 metres, or 2.5 times the length of the same group's smallest car, the Smart Fortwo.

Child star Jackie Coogan owned two Rolls-Royces before he was 21, one of them adorned with a radiator mascot of himself playing 'The Kid' opposite Charlie Chaplin. Budding English actress Diana Fluck changed her name to Diana Dors and somehow managed to buy a Silver Lady-adorned limousine on the never-never when penniless and 20, demonstrating more faith in her future than anyone else.

SIMPLY THE BEST

Phantom of the opulent: the long wheelbase version of the first 'BMW' Roller. Buyers were offered a palette of 45,000 external colours but subtlety wasn't on the options list.

And how much did the Maybach cost? As they say, if you have to ask . . . you probably earned your money through honest work.

Rolls-Royce followed its original 'best car in the world', the Silver Ghost, with such models as the Silver Wraith, Silver Dawn, Silver Cloud and Silver Shadow. And it built near-clones with Bentley badges and Bentley grilles for people who wanted to stick it up the neighbours a little more quietly.

Inside the Rolls-Royce Phantom, showing the finest British craftsmanship that Germany can offer.

Most of the competitors at the very top end had fallen away by then, and Rolls nearly followed them in 1971. The problem was the Rolls aircraft division, which had shown that just because you can build complex jet engines doesn't mean you can add up. By massively underquoting on the

---

When they arrived in Sydney in 1964, The Beatles were picked up from the airport in official 'Beatle Limousines'. Rolls-Royces? Daimlers? Cadillacs? No, Morris 1100s.

211

> 'Forty years ago, cars were designed to be a combination of a rocket and a woman . . . cars were the canvas on which the American dream was painted. Cars used to mean escape, speed, power. Now, it is gas mileage and air bags. [They] have gone from a metaphor for freedom to a metaphor for our own mortality.'
>
> Robert Thompson, a professor of pop culture - yes, they do exist - at the US's Syracuse University, goes out of his way to spread good cheer (1999).

development of a new engine for the Lockheed Tristar, it sent the whole organisation bankrupt. Only nationalisation by the British Government kept the show on the road.

The car-building portion of Rolls-Royce was spun off separately in 1973 and re-privatised, but for the rest of the century sales were painfully slow and adoption of new technology by the brand was even slower still. Despite this, surveys regularly showed that Rolls-Royce still remained the 'best car in the world' to more people than any other. Products as diverse as computers and running shoes continued to be spruiked by their makers as 'the Rolls-Royce' of their category, even if they weren't ludicrously overpriced, dated in their styling and filled with 1950s technology.

During the 1990s most of the world's independent car makers were snapped up by the big boys and Rolls-Royce was no exception. The Volkswagen Group stumped up £430 million (about Aus$1.2 billion) for 'Britain's finest' in 1998. It was a ludicrously high price considering Rolls-Royce's negligible earnings, and a particularly ironic purchase for a company named in honour of the people's car. But there was something even stranger to follow. In its rush to outbid BMW, which was also keen to secure the famous British marque, VW failed to read the small print. Its executives discovered that although they had bought the famous factory at Crewe, and the right to use the Spirit of Ecstasy symbol and the famous radiator grille, there were two words it didn't have the right to use: Rolls and Royce. VW could use

> Chrysler launched an upmarket version of the Imperial, designated FS, in the US in the 1970s. The FS stood for 'Frank Sinatra' and the price of the car included 16 cassettes of you-know-who with eyes of blue. The car came only in black, and no doubt could fit a decent-sized violin case in the boot.

the Bentley brand-name if it wished – that was owned solely by the car division – but the Rolls-Royce moniker was owned by the independent aero division and VW couldn't use even the hyphen without a separate leasing deal.

BMW jumped back into the ring and, for the relative bargain of £40 million, grabbed the right to use the name Rolls-Royce on cars. It looked like a clever move, except that BMW couldn't use the silver lady or the trademarked radiator. A deal between the two German companies was eventually (and painfully) nutted out and from 2002 BMW ended up with the Rolls-Royce brand, lock, stock and walnut burr barrel. VW kept the legendary Crewe factory (that's legendary as in 'old' and 'past its best') plus the Bentley name.

'We only wanted Bentley in the first place,' said VW executives, who genuinely seemed to think someone would believe them.

There was widespread carping at the idea of a German Rolls-Royce, but in 2003 the first BMW designed-and-built Roller, known as the Phantom, made a bold attempt to return

> **'Vorsprung durch Technik.'**
>
> Audi's German language slogan (meaning 'progress through technology') became a buzz-phrase in the 1990s and surfaced in the title track lyrics of two of that decade's biggest selling albums: U2's *Zooropa* and Blur's *Parklife*.

That peculiar US luxury vehicle, the 'stretch limo', has inspired some strange and silly machines. This is the 6 metre-long Mini XXL, with two extra seats, two extra doors, two extra wheels, and a spa bath in the rear.

the marque to its roots. Which is to say to as a car so large, luxurious and fantastically well equipped that only the ultra-rich need apply.

The previous British-designed Rolls-Royce Silver Seraph could be easily lost in the traffic, but now it was the traffic's turn. From any angle the enormous V12-powered Phantom looked somewhere between imposing and bombastic. The bonnet was at waist height even for a tall man, the roof line at iris level. The bonnet went on forever and the traditional Rolls-Royce grille was of such a size that even a gold-chain-draped rap artist may have considered it a little ostentatious.

The driving position was modelled on the bridge of a boat to give what the designers called an 'authority driving position', the sound system had nine amplifiers and 15 speakers, while sumptuous and exotic materials covered every interior surface. And the Australian price at launch was $985,000, the equivalent of 66 Toyota Echos. Like it or hate it – and there were plenty in the second camp – Rolls-Royce had returned with a bang.

THE FIVE GREATEST CARS OF ALL – A LIST TO ARGUE OVER

Ford Model T (1908)

Henry Ford was a nutter and not always a benign one. His greatest creation carried over many of his quirks and foibles: it was eccentric, it was austere, it had fascist leanings and a fascination with square dancing (OK, to be fair the last couple of points applied only to the man).

Yet the four-cylinder, 15 kW Model T was also light, tough and reliable. It was created by a man who, for all his flaws, genuinely believed an affordable car could change

> 'Someone should write an erudite essay on the moral, physical, and aesthetic effect of the Model T Ford on the American nation. Two generations of Americans know more about the Ford coil than the clitoris, about the planetary system of gears than the solar system of stars.'
>
> Nobel Prize-winning American author John Steinbeck in *Cannery Row* (1945). Unfortunately Steinbeck never took up his own suggestion and wrote said essay.

All things to all men (and women): The T-Ford.

the lives of working folk for the better. He believed he had a mission to supply it.

The greatness of the Tin Lizzie also rests with the way it was built. It was screwed together on a moving production line, it had completely interchangeable parts and every process, from sorting of raw materials through to painting (only in quick-dry black from 1915 to 1926), was devised to maximise efficiency and minimise the purchase price.

You didn't need a colour other than black if the alternative was walking. And for many it was. Those who couldn't afford a Flivver this year could wait for next year, at which point it wouldn't have new headlights, revised trim patterns and a 'Mark 3' badge. It would be pretty well exactly the same, except cheaper. And a T could be adapted into anything: a race-car, a hearse or a tractor.

In 1918 Ford built 642,750 Model Ts, single-handedly accounting for 50 per cent of the world's entire car production for the year. It proved that Henry had devised exactly the right package for the era. He didn't know when to stop – by the 1920s the T was lagging behind the competition in almost every area except price and durability – but the point had been made and a true people's car given breath.

The final tally exceeded 15 million, a figure that wouldn't be broken until a certain German machine beetled on past the Model T's total in the 1970s.

VW Beetle (1938)

It was launched in Germany by, er, a politician, and production was almost immediately put on hold in favour of another of the same man's projects: World War II. Yet a decade or so later the rugged little rear-engined two-door machine started criss-crossing the world and delighting millions. This happened because the VW was everything the politician who conceived it wasn't: cheerful, honest, humble, egalitarian and air-cooled.

The Bug has inspired books, sculptures, paintings, even classical music (Harry Phillips produced a Concerto for Yellow Volkswagen and Orchestra, with the score incorporating the sound of the engine, horn and slamming doors). It connects with people in a way few vehicles do. Has anyone ever proudly worn a Kia necktie or tried

VW prototypes from before World War II.

And a 1952 'split window' production model.

cramming students into a Mazda? Would Walt Disney make a series of films about a magical Ford Laser? Can you imagine, 30 years out, people buying Hyundais and lovingly restoring them because their youth just wouldn't have been the same without the Accent 1.6?

Although German production stopped in 1978, until 2003 they were still building the Vee-Dub in other countries to something close to the original recipe, and by then 21,529,464 had been produced. Sure, every one of those 21,529,464 examples was a little cramped, slow and unstable. But these are small prices to pay for mechanical simplicity, design purity and giving a solid poke in the eye to the fashionistas.

Beetles are noisy, but the clever placement of that tough little flat four means you are driving away from the racket rather than into it. Anyway, even the sound is iconic: the Dak-Dak nickname identifies the only car in the world the average punter can pick by its engine note. It is also a machine almost anyone can immediately recognise by its shape. And like the T before it, a Bug can be easily

> 'The vehicle does not meet the fundamental technical requirements of a motor car. As regards performance and design it is quite unattractive . . . too ugly and too noisy.'
>
> A British motor industry commission, headed by Lord Rootes (of Hillman fame), assesses the merits of the VW people's car at the end of World War II. The report added that the odd-looking German car could never be mass-produced cost-effectively, and advised that neither the plant nor the design was worth anything to Britain as a war reparation. The 20 millionth Beetle was built in 1981. By that time the British industry was in tatters.

converted into anything you want it to be: a beach buggy, a sports car or an amphibian. And no matter what it ends up as, you can't kill it with an axe. We shall not see its like again.

Citroën DS (1955)

The DS is the most lauded French car of all, nominated by various panels and juries as 'car of the twentieth century', revered by Jackie Kennedy and scores of movie stars, and credited with saving the life of President de Gaulle during a 1962 assassination attempt.

And while you can argue about which car was most ahead of its time, no mainstream model has ever *looked* more ahead than the Citroën DS.

Fit for a goddess: the elegant, wildly original Citroën DS.

When it appeared at the Paris Motor Salon of 1955, the DS's stunningly original silhouette and gobsmacking technical specifications, including hydropneumatic self-levelling suspension, made everything else seem mundane. The punning name – *déesse* is French for goddess – also hit the spot and a reputed 79,000 orders were taken at the Salon alone.

At a time when Holden was trying to convince Australians the 'humpy' FJ sedan was thoroughly modern and most American manufacturers were merely tweaking the sheetmetal each year, the big Cit may as well have fallen out of the sky.

Citroën's PR department claimed that only the death of Stalin two years earlier had generated as much post-World War II newsprint. French intellectual Roland Barthes even produced a philosophic treatise on the DS (something he inexplicably failed to repeat when Holden launched its FE model the following year).

'The DS has all the features,' Barthes wrote '. . . of one of those objects from another universe which have supplied fuel for the neomania of the eighteenth century and that of our own science-fiction: the *Déesse* is first and foremost a new Nautilus . . .

'It is possible [it] marks a change in the mythology of cars. Until now, the ultimate in cars belonged rather to the bestiary of power; here it becomes at once more spiritual and more object-like . . .'

If you didn't consider the standard model sufficiently spiritual or object-like, there was the even more desirable two-door convertible version, delightfully named Decapotable.

> **'One of the biggest advances in production car design in the whole history of motoring.'**
>
> A typical comment on the Citroën DS, introduced in 1955. This one was from Laurence Pomeroy, technical editor of Britain's *The Motor*. Journalist/racer Paul Frere said 'it gives the impression of having jumped a generation in automobile history'.

STEP ON IT!

> **'Better in jams than strawberries.'**
>
> One of the many clever slogans used to promote the Mini in England (this one from the mid-1970s).

Mini-Minor (1959)

Why increase weight and reduce usable space by driving the rear wheels when there are two perfectly good wheels up front next to the engine? Why have a long bonnet when you can turn the engine sideways? Why bother with bulky and heavy steel springs for the suspension when you can use cones of rubber? Why add cost and time-consuming production processes to hide the welds on the body seams or conceal the door hinges?

Alec Issigonis's front-drive Mini, or 'brick', treated all these questions as rhetorical. It was a masterpiece of

Nimble, sparse and capable. So capable, in fact, the Mini won the Monte Carlo Rally in 1964, 1965 and 1967. It came first, second and third in 1966 too, but was disqualified on a dubious technicality by the French judges, who awarded victory to, you guessed it, a Citroën.

original thinking. It was the biggest thing in small cars ever, so clever the shape still makes perfect sense in reborn form (while that of the space-inefficient New Beetle doesn't). Sure, the Mini had some failings, such as an annoying offset steering wheel, but it crammed in more virtues than had ever been fitted in such a small mechanical package.

That its roadholding could be more impressive than most big cars seemed to completely defy conventional wisdom. Yet the Mini ran rings around even some supposed sports machines, thanks to firm springing and having its four wheels pushed out to its very extremities.

Due to appalling accounting, the company that built it – BMC – lost money on every one built during the first few years. It also produced entirely stupid variants such as the Wolseley Hornet and Riley Elf. Yet along the way it gave us the sublime blending of economy model and racing machine that was the Mini-Cooper. The world had seen micro cars before, of course, but until the Cooper none of them had won major races and rallies.

Almost every successful small car since the 1960s owes a debt to the original, fabulous brick. And surveys have regularly shown that Mini fans are equally male and female, wealthy and hard-pressed, flamboyant and conservative. If any car is the metallic embodiment of good cheer and equality, this is it.

With around 90,000 residents and more than 100,000 road vehicles, the Channel Island tax haven of Jersey has the highest per capita car ownership in the world. The island's speed limit is 40 mph (approximately 64 km/h), while many 'Green Lanes' restrict vehicles to 15 mph (24 km/h). That must account for all the Porsches and Ferraris.

STEP ON IT!

Lamborghini Miura (1966)

There is absolutely no logical reason to list the Miura here. It was but one of many noteworthy Italian supercars of the 1960s and 1970s. It didn't necessary do anything better than its competitors and had a habit of catching fire.

Yet that is exactly why it has been included. Cars aren't fridges or lawn-mowers; they can't be assessed completely objectively because they have personality and style, and personality and style are entirely in the eye of the beholder. When everything comes together correctly, cars have an ability to do something mere appliances can't: they can move our soul. And the Lamborghini Miura can move a soul at better than 280 km/h to the aural accompaniment of a 12-cylinder, 3.9-litre double overhead camshaft engine capable of spinning at nearly 8000 rpm.

It first moved this writer's soul in primary school. I had a Matchbox or Dinky version and fell in love with it, convinced that no-one ever had, or ever could, design a machine quite as beautiful. I was entranced by the way the

The Lamborghini Miura.

Official opening: the Miura's bonnet and boot were works of art.

whole back of the Miura lifted up to reveal the sensationally complex transverse V12 (which was placed amidships) and the front bodywork lifted up in almost perfect symmetry. I was captivated by the upswept doors and how, when both open, they looked like the bull's horns in the company's logo. The sharklike nose, the strange flutes around the tilting headlights, the race-car-style cockpit . . . it was all so exciting I sat down and memorised not only the car's vital statistics but learned how to spell Lamborghini, the longest and most exotic word I had encountered up to that time.

Everyone with blood in their veins should have a Miura equivalent in their heart. If they ever really did own one, it may well disappoint (or, in the Miura's case, immolate), yet as an automotive ideal it perfectly reminds us of everything a great car can be – and a fridge, lawn-mower, computer or almost any other machine can not.

> **'If you don't like the way I build my cars, why don't you build your own in that tractor factory of yours?'**
>
> The reputed response of Enzo Ferrari when the wealthy industrialist Ferruccio Lamborghini complained that his own Ferrari had some design flaws. Another version has Ferrari, Italy's 'Pope of the North', refusing to meet Lamborghini, saying 'A tractor manufacturer could never be expected to understand high-bred sports cars.'

Lamborghini has long favoured bull-fighting names for its cars. The Murciélago is named after a bull that put up such a fight in the ring in 1879 that its life was spared. Murciélago was given for breeding to Don Antonio Miura, whose surname ended up on what many believe is the best Lamborghini of all.

Traditionally books covering the history of the car finish with a photo of an experimental car meant to look futuristic yet which already looks slightly dated by the time the book is in the stores. Fortunately the strongest trend in the past decade is to build concept cars that look like things that have gone before. So here instead is the 2006 Lamborghini Miura, capturing much of the magnificence of the original and probably much less flammable and, below, the Audi Rosemeyer, a spectacular showcar from 2000 designed to recall the Auto Union 'silver arrows' of the 1930s.